Loggers and Lumbermen

The Evolution of the Forest Industry in the Southern Interior of British Columbia

This book is a project of the

Interior Lumber Manufacturers' Association

1994

First published by
ILMA
Interior Lumber Manufacturers' Association
#360, 1855 Kirschner Road
Kelowna, B.C. V1Y 4N7 Canada
1994

Canadian Cataloguing in Publication Data
Main Entry under Title:

Halleran, Mike, 1935-
LOGGERS AND LUMBERMEN

Includes index.
ISBN 0–9698187–0–X

1. Forests and forestry–British Columbia–History. I. Interior Lumber Manufacturers' Association. I. Title.
SD146.B7H34 1994 634.9'09711 C94-910276-8

Design and Printing by
Ehmann Printing Limited
Kelowna, B.C. Canada

Using recycled paper

Author's note

In researching and compiling the data that became the foundation of this book, Fred Stinson recorded the evolution of the industry through the geographic regions of B.C.'s southern interior – the ILMA's operating area. It became apparent, while transposing the Stinson research material into book form, that these geographic operating regions coincided closely with the biological, geological and climatic zones that created the various forest types in the first place. Historically, and currently, these land forms and climatic conditions have distinguished the forest types and the nature of forestry practised in the southern interior.

From the desert dryness of the south Okanagan to the wet valley bottoms of the West Kootenay, varied conditions have created a natural diversity of plants and animals. These "bio-geo-climatic zones" are complex and unique environments. The varied natural conditions have not only determined the tree species, but also the logging methods and the type and location of road construction. Climate and terrain have even defined the seasons in which work could progress.

From the beginning – and sometimes the hard way – people in the forest industry learned to adjust operating choices to the conditions and values that existed on given sites. Industry operations were also governed by prevailing social and economic conditions at given points in time. Failure to take any of these variables into account could set the stage, not only for economic difficulty, but also for environmental and community disruption.

This recounting of forest industry history spans the period from pioneer days to the present. Forestry is much changed from early times. It seems safe to anticipate that change will continue. It is the writer's hope that there will be more universal awareness that change is both natural and necessary. Perhaps then more favorable light will be shed upon the forest industry as a sustainable and valued presence in the southern interior of British Columbia.

MIKE HALLERAN
Meadow Creek, B.C.
Spring, 1994

The Interior Lumber Manufacturers' Association publicly acknowledges the invaluable contribution made to this book by Fred Stinson. Compiling the historical data on the hundreds of mills that once operated in the southern interior was a monumental task. Completing this project would not have been possible without all the useful information provided by Mr. Stinson.

The ILMA publicly thanks Finning Ltd. for its generous
financial assistance toward this project,

&

The ILMA extends its thanks to all those who contributed information concerning their role in the forest industry of the southern interior of B.C. Many provided fascinating personal anecdotes, information on dates and events, and above all photographs (many of them family treasures). Their contributions were of enormous help in bringing this project together.

The ILMA also thanks members of the History Book Committee:
Don Fraser, Barney Lukas, Al Macdonald,
John Marritt RPF, John Murray RPF,
Gerald Raboch, Fred Stinson RPF, Bill Uphill and Theresa White.

Dedication

The vision for celebrating the pioneering spirit of loggers and lumbermen in the Southern Interior grew as the ILMA approached its 50th anniversary. Now three years beyond that milestone, the process of gathering historical records and anecdotes from the corners of the Southern Interior has managed to develop quite a history of its own.

Gathering and weaving the strands of forestry's evolution has admittedly been a massive undertaking, but definitely worth the investment of time. We now hold in the highest regard those whose energy and creativity has passed on a precious forest legacy to the communities of the Southern Interior. These are the pioneers and trailblazers to whom this book is respectfully dedicated. May we, in turn, pass on as rich a legacy to the people of the 21st Century.

Contents

Cedar stand – Big Bend, Columbia River, 1938.

Introduction

The mountain ranges of the western cordillera, all of them predecessors of the Rocky Mountains, define the various geographic regions and ecological conditions of the southern interior. These mountain chains create natural boundaries that have ultimately determined the nature and type of the forest and, consequently, the type of forest operations.

These mountains act as "walls", natural separations between the different biological, geological and climatic zones which make up the most diverse set of ecosystems in B.C.

A Region of Great Natural Diversity
(From Temperate Rain Forest to Desert)

The natural diversity of the southern interior region is profound. The cedar/hemlock valley bottoms in the Purcell Trench are similar to coastal vegetation types. Annual precipitation (combined rainfall and snow) in the Purcell Trench is close to 80 inches per year – more precipitation than in the western end of the Fraser Delta. By contrast, in the Rocky Mountain Trench, one mountain range to the east, annual precipitation is about one-quarter as much. There are few, if any, cedar or hemlock trees growing there, but plenty of aspen and sage. Interior Douglas fir tends to dominate in the lower elevations of the Rocky Mountain Trench, along with lodgepole and yellow pine. There are even a few examples of open grassland.

A still drier environment is encountered in the southern end of the Okanagan Valley – between Osoyoos and the American border. This area, receiving less than 10 inches of precipitation per year, qualifies as a true desert.

The diversity of these natural "biogeoclimatic" zones determines more than vegetation and animal types. It also determines the kinds of forestry practices that are most suitable and sustainable. Obvious examples are the (cedar) shake and shingle operations still being carried on in the valley bottoms of the Selkirk and Monashee ranges. Only on B.C.'s west coast do cedar trees grow larger than those in the interior wet belt of the West Kootenay.

Early Reforestation (Left Largely to Nature)

In recent years foresters, biologists, geologists and other specialists have studied these various regions extensively to determine the best land use* activities. Their studies are not simply academic exercises. Many are driven by hard economic need.

An excellent example relates to reforestation. Those familiar with early reforestation efforts in the southern interior are aware that plantation forests were not exactly unqualified successes. Some were bitter disappointments. Many reforestation problems were caused by planting stock which originated from other regions with quite different growing conditions. In some cases, natural regeneration (after logging or fire) progressed more quickly and effectively than replanted sites.

More recently, with increasing awareness of diverse natural conditions, reforestation efforts have met with much more success. Seed taken from a given site is returned as a developed seedling to the same area. What seems an infinitely logical and natural choice, was not recognized in the early days of B.C. nurseries and reforestation. There are now "regional" nurseries, producing genetically improved stock that is returned to natural sites that provide better expectations of success. Records show that today's reforestation programs are more extensive, and far more successful.

It took approximately 50 years to plant the first billion seedlings in B.C. The second billion took only eight years and the third billion only four years. According to the Forest Service, planting survival now exceeds 80 percent.

Lots of Common Sense – Not Much Science

To some it may have appeared that the industry and the entire forest community were slow-learners. In reality, the first 50 years of forest sector activity in B.C. was extremely non-technical in nature. Few technical or scientific skills were involved – nor were they needed. In the early days the challenge had more to do with straightforward, but physically demanding tasks – building roads, cutting down and hauling trees, milling trees into lumber and delivering wood products to market.

During the 1960s – barely 30 years ago – the forest sector began to feel more pressure to accommodate other land uses. Sometimes other interests were in head-on competition for the same piece of countryside. Gone with the first half of the century were the days when an operator (and perhaps an assistant ranger) sized up the ground, picked out the best place to start, fired up the "Cat" and started logging. Now, before a road can be built or a tree cut, it is necessary to work through a maze of regulations and requirements so confounding in nature as to make overhauling a transmission look simple.

If forestry before the 60s was less regulated and more easy-going, all that has now changed forever. The natural diversity of the southern interior, comprised of that great complexity of "biogeoclimatic zones", added to complications. Operating methods deemed appropriate in one setting, could prove to be unworkable, uneconomic or environmentally unsuitable on another site. Thus site-specific forest management planning became the rule, rather than the exception.

From the beginning, loggers and sawmillers would have been aware of the southern interior's natural diversity. They would have observed how much conditions could change within the radius of a few miles. This history reveals how early forest companies first designed their operations around economic opportunities and pitfalls. It also shows how these companies have adapted to address the environmental challenges of operating in diverse and sensitive natural conditions.

From the beginning, constant change

At about the turn of the century the demand for lumber, railroad ties and other forest products created unparalleled opportunity for a pioneer forest industry in the southern interior. Ninety years ago, without infrastructure or mechanization, early industry followed the building of the railroads into the valleys of the interior. There were at least two rail lines in place by the early 1900s, and later some of the bigger forest companies built their own railroads.

"Some of the forest companies built their own railroads."

Steamboat service, available on larger lakes and rivers, was about the full extent of mechanized transportation. When logging and sawmilling activities began in the southern interior, the great network of main and secondary roads was still decades in the future. Lack of roads wasn't a problem, because logging trucks hadn't been invented yet.

The first mills were steam-powered and wood-fired. Steam had some definite advantages. For one thing, as the mill cut lumber from logs it also created its own supply of fuel. The most efficient steam mills burned slabs and hog fuel (bark and sawdust) produced in the mill itself. In some cases, the steam plants also powered electric generators. Where mills were located on a lakeshore and fed from log booms, underwater steam lines kept the booms from freezing in the winter. Steam-powered mills were also quiet. From even a short distance away, saws could be heard but the power plant itself was virtually silent.

The steam mills were nearly all phased out by the late 1940s. Many old-time sawmillers resisted the conversion of mills to diesel power (diesel mills were terribly noisy by comparison). But nothing could match the efficiency of diesel power for portable mills and there were, at one time, literally hundreds of portable mills. When diesel power was eventually converted to electricity, mills were even quieter than in the days of steam.

"The first mills were steam powered and wood-fired." Yahk Lumber Company.

"Trees were felled with axes and hand saws – usually two-man crosscut saws." Adams River Lumber Co. (Photo courtesy Kamloops Museum)

Real horsepower moved logs in the early days.

Lumberton

Logs were skidded with horses and there were a few wooden spar trees. On some winter shows, logs would be loaded on bob-sleighs and road-hauled to mills or railroads.

"When I came up to the interior they were still logging with horses. The teamsters would all go down to the horse barn after supper and stand around and be with the horses and talk, have a smoke and sort of wind down from the day. They would pet the horses and give them a little treat or some feed. It was a very nice feeling and I never forgot it. I was kind of sorry to see the horses go. But they could not operate in the steeper ground and we had to go to machines."

– FRED STINSON, (ILMA PRESIDENT 1977-79)

"I started working with my father when I was in my teens. I had to learn the techniques of working with horses. One of the things I had to learn was respect for the horses. I gave them a tremendous amount of care and I looked upon them as comrades and friends more than as slaves. We had a small company with eight horses at one time. They had a very precise time clock. They knew at 4 o'clock in the morning it was time to be fed and watered. They also knew when it was 5:00 p.m. and they would just quit working; no ifs, ands or buts about it."

– GERALD RABOCH, (CHIEF OPERATING OFFICER, RIVERSIDE FOREST PRODUCTS LTD.)

Logs delivered by chute for Adams River Lumber Co., early 1900s. (Courtesy Kamloops Museum)

"We were hauling logs with teams and sleighs. To unload the logs you had to hook your peavey (a lever, with steel hook and point, used to handle logs) under the bottom log and pry it loose and then the whole load would tumble off. Well, I lost my peavey under the load one time and of course the handle got broken. So on the weekend (on my own time) I had to go out and find a yew tree and cut it down and make a new handle. That was my punishment. That would be back in 1933 and peavey handles were pretty expensive at that time."

– IAN MCDIARMID, (ILMA PRESIDENT 1964-65)

Bull River

Rivers and other waterways became transportation routes for logs. Log drives were quite common on the rivers of the southern interior until better roads, better trucking and more concern for fish habitat stopped them. Some companies built and operated elaborate flume systems. Water courses were contained in wooden flumes large enough to float round logs to a sawmill. Some of these flumes were miles long. They were expensive to construct, but since the energy source – flowing water – was self- renewing, there was no ongoing transportation cost to the operation.

Lumberton

Lumberton log flume.

Lacking mechanization, early logging operators used flowing water to transport logs wherever possible. Log drives were common on interior rivers in the early days and early loggers, being talented and innovative, built elaborate flume systems to transport logs. Some of the flumes were marvels of design and engineering.

"In the early days, logs were driven down rivers to the mills. Horses were used to bring the logs to the river bank and the spring freshets brought them down. In the early 1920s there was a fairly large sawmill at Enderby. The logs were driven down Sugar Lake to Shuswap, then to Mabel Lake and finally via the Shuswap River to Enderby. When the drive was completed you could walk on logs from Enderby four miles upstream. The salmon survived. They are still there."

– GERALD RABOCH, (CHIEF OPERATING OFFICER, RIVERSIDE FOREST PRODUCTS LTD.)

Most mills started small and some never did get to be very large. Following the Second World War, the industry experienced a period of unprecedented growth. Lumber demand responded to the "baby boom" and increased housing starts. For as long as they were able to secure a good log supply, many of the smaller mills prospered. But as the most accessible wood was exhausted, and operators had to reach further and further back, it became more difficult for small mills to remain viable.

Traditionally innovative and self-reliant, many small operators were able to hang on by relocating their mills closer to the timber supply. These moves heralded the era of the "portables" and bought more time for many operations. But by the late 1950s, dozens of mills started to drop out of the industry. The great period of consolidation began.

The following history of mills throughout the southern interior verifies these trends. Smaller mills sold to larger, better financed operations. The small mills were often bought mainly for their timber supply. Many were bought and immediately closed down as centralized, high capacity, modern mills became the dominant force in the industry.

The development of the forest industry is reflected in the development of interior communities. Forestry proved to be so powerful an economic instrument that it figured prominently in the growth and stabilization of the entire region. In town after town in the B.C. southern interior, forestry made it happen.

Regional Histories – Introduction

Explorers and pathfinders were the first to arrive. Then came independent trappers, and following on their heels, agents and traders for the large fur companies. Gold seekers and miners usually came next. As soon as there was enough settlement to create a market for beef, and sometimes even before that, pioneer ranchers had cows on western ranges. Most of the towns that sprang up needed some kind of sawmill. The towns were located near water, which was convenient, as mills were either driven directly, by water, or indirectly, by steam.

At first, these mills were exclusively community operations that supplied local markets. Sales were linked to the local resource industries that attracted people to settle in or near the town. The timber industry of the B.C. interior, as in many other parts of the west, began as a locally based service industry. In the latter half of the century, the local focus was destined to change gradually, but radically.

With the continent-wide building boom that followed the Second World War, export sales rose rapidly. Now, the majority of B.C. lumber is produced for export. On the coast, the export of lumber and other wood products got started earlier and for obvious reasons. It was easy to ship lumber overseas when high quality trees were standing within skidding distance of tidewater. But interior forests had to wait for a transportation network and an expanding demand before the shift from essentially supplying local markets to serving an export market.

Railroad ties – One of the first forest products

In many parts of the interior, railroad ties became the first commercial forest product. Wherever railroads were built, a nearly insatiable demand for ties was created. Railroads use more than 3,000 ties per mile. Tie-cutting became big business from the moment the railroad boom began.

These thousands of ties were made by hand. After the trees were cut into lengths with axes and either swede or crosscut saws, the lengths were "hewn" to make a finished tie. Tie "hackers", some carrying out their craft even into the 40s and 50s, usually scored the sides of the round log with a conventional poleaxe (single bitted). Then a broad axe – an instrument flat on one side and angled on the other – was used to "hew" the ties flush along the score-line. The broad axe handle was offset from centre a couple of inches so the axeman's hands would be clear of the work. The handle could be pounded out and set in from the opposite side in case the next axeman was left handed; or wanted to switch arms. Handy.

Lumberton

Some enormous hand-hewn timbers occasionally found in old buildings and bridges display phenomenal precision. In the West Kootenay, hand-hewn bridge timbers of two-foot square dimensions, more than 30 feet long and of perfect uniformity, have been noted. Even into the 70s and 80s, during routine railroad maintenance or while salvaging rails from abandoned branch lines, section hands were still finding the odd hand-hewn tie. Claims have been made that hand-hewn ties have longer life spans than sawn ties – an unprovable assertion given the number of variables that affect a tie's life. Some of these variables are soil types, ground moisture and tree species from which ties are cut. The broad axe did leave a smoother surface, much like the difference between rough and planed lumber. Some have suggested that this glazed effect helped ties resist water penetration better than sawn ties.

Transportation – Moving the Wood

Canadian Pacific, Canadian National, Great Northern and a number of smaller carriers rushed to get rights-of-way into the valleys of the interior. These large rail

Railroad building provided one of the first and most reliable markets for forest products. As many as 3,000 ties were required for each mile of track.

companies were preceded by pioneer logging companies that had laid their own track to haul logs and lumber to meet established commercial carriers. Some of these long-abandoned logging railroads, though overgrown and forgotten, can still be traced in parts of the interior.

As soon as transportation ceased to be a problem, the rich forests of the interior were able to contribute more and more to the province's export trade. Interior forests were almost indescribably vast and were widely regarded as inexhaustible by the pioneers. It is no wonder that a myth of superabundance prevailed.

The coastal forest industry cut more volume than the interior from the very beginning. But gradually, as demand grew and access expanded, their positions began to reverse. The interior harvest has exceeded the volume logged on the coast for many years.

Regretfully, few truly high quality pictures of 19th century southern interior logging appear to have survived. Even from the early part of this century, there are few good pictures. As this history unfolds, the usable photographic records we have been able to gather – enhanced through computer technology – should provide a sense of the place, the people and the forest activity of the time.

The old logging railroads exemplified the resourcefulness of early loggers. They used wood from the site for ties and bridges. In some cases, even the locomotives were wood-burning, powered with material from the logging operation.

Early years (south of Cranbrook).

"Today you see a very large percentage of forest products moved by truck. It seems that before, everything moved by rail. Chips were moved by rail and lumber too. I have seen the abandonment of the Boundary CPR sub-division from Castlegar through to Midway and the mode of transportation had to be trucks. There has been a very definite change from rail to truck transportation and I see that trend continuing."

– ABE FRIESEN, (PRESIDENT, POPE & TALBOT LTD.)

Early logging south of Cranbrook.

1922 – Near Fort Steele

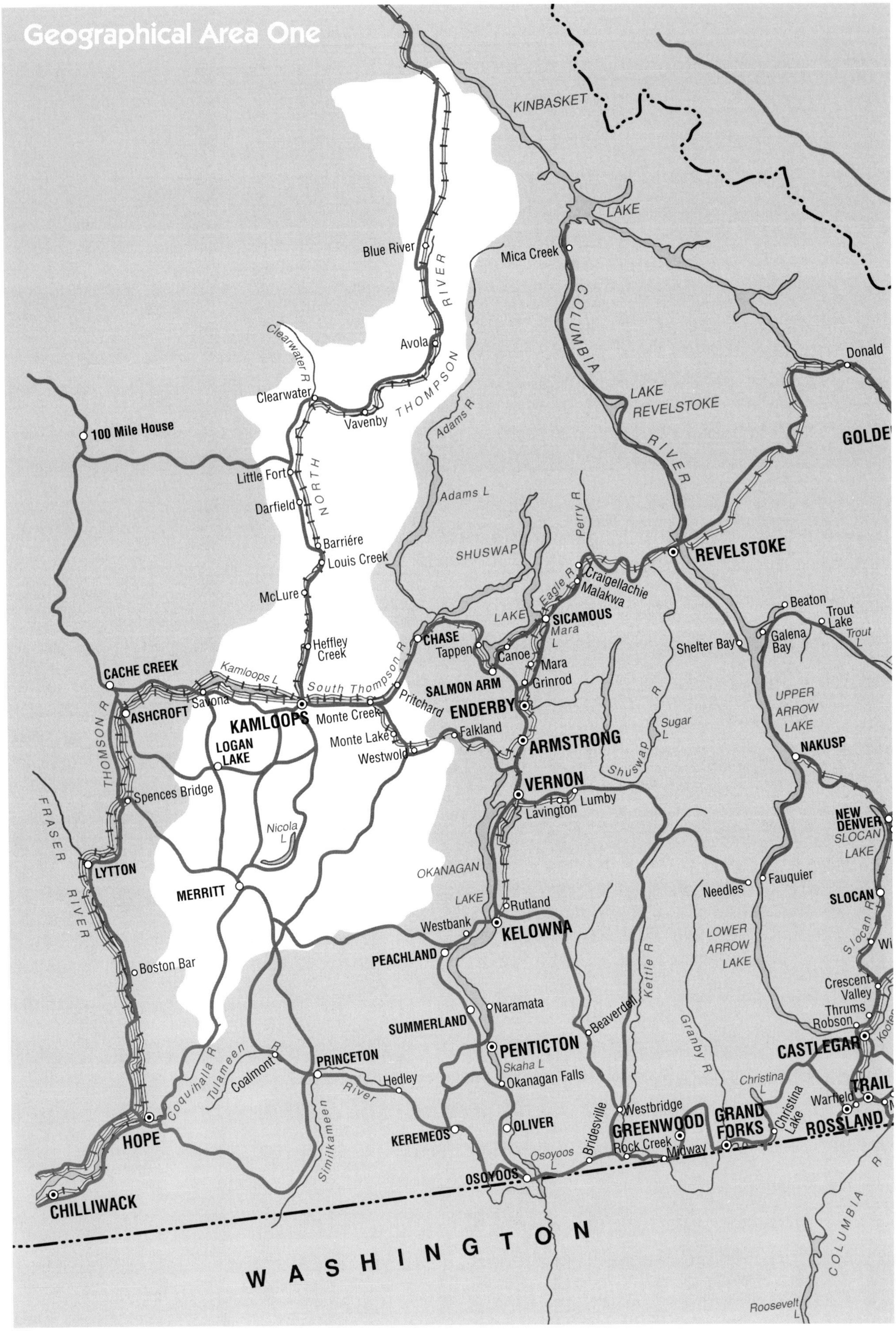
Geographical Area One
KINBASKET
LAKE
Blue River
NORTH THOMPSON RIVER
Mica Creek
COLUMBIA
Clearwater R
Avola
Clearwater
Vavenby
Adams R
LAKE REVELSTOKE
RIVER
Donald
GOLDEN
100 Mile House
Little Fort
Darfield
Adams L
Barriére
Louis Creek
SHUSWAP
Perry R
REVELSTOKE
McLure
Eagle R
Craigellachie
Malakwa
LAKE
SICAMOUS
Mara L
Beaton
Trout Lake
Galena Bay
Trout L
Heffley Creek
CHASE
Tappen
Canoe
Mara
Shelter Bay
CACHE CREEK
Kamloops L
South Thompson R
Grinrod
SALMON ARM
Pritchard
ENDERBY
ASHCROFT
Savona
KAMLOOPS
Monte Creek
UPPER ARROW LAKE
THOMSON R
LOGAN LAKE
Monte Lake
Westwold
Falkland
ARMSTRONG
Sugar L
Shuswap R
NAKUSP
Spences Bridge
VERNON
Lavington
Lumby
FRASER RIVER
Nicola L
NEW DENVER
SLOCAN LAKE
LYTTON
MERRITT
OKANAGAN LAKE
Rutland
Needles
Fauquier
SLOCAN
Slocan R
Westbank
KELOWNA
LOWER ARROW LAKE
PEACHLAND
Boston Bar
Kettle R
Crescent Valley
Thrums
Robson
Naramata
SUMMERLAND
Beaverdell
Granby R
Coquihalla R
Tulameen R
Coalmont
PENTICTON
Skaha L
Okanagan Falls
CASTLEGAR
PRINCETON
Hedley
River
Christina L
Christina Lake
TRAIL
Warfield
Similkameen R
HOPE
KEREMEOS
OLIVER
Bridesville
Westbridge
GREENWOOD
GRAND FORKS
ROSSLAND
Rock Creek
Midway
Osoyoos L
OSOYOOS
CHILLIWACK
COLUMBIA R
WASHINGTON
Roosevelt L

12

Geographical Area One
The Thompson Basin Region

Word As Bond:
Ken Long and Hugh Leir

Ken Long is the sole surviving founding member of the ILMA.

> *"In my entire lifetime in the mill business everything was done with a handshake and our word. It was done by maintaining goodwill between the parties. We went through nearly 40 years and never had a contract with anybody."*
>
> – Ken Long

In the Merritt country in 1945 two sawmill operators, Ken Long and Hugh Leir, shook hands on a deal. They promised to conduct their competing operations within certain areas of the region so as not to interfere with each other. Long bought Leir's sawmill site near Merritt (now occupied by Tolko Industries) for $4,000. Leir took up operations just over the ridge from him, on the Princeton side of the mountain. This gentlemen's agreement was apparently respected by both parties without the benefit of written language. There were a lot of "handshake" agreements in those days. Some of them ended happily – others not so.

"In 1957 we bought McLean's Sawmill at a place called Malakwa. All the logging was done by contractors. When we moved in there the first thing the loggers wanted to see was a written contract. They could not believe it when we told them that we worked with our contractors without the need for written contracts. So we gave them the names of other contractors who had been successful working for us. We said they could talk to any one of them, and they did. I think that was a great period and it was a great way of operating."

– HORACE SIMPSON, (RETIRED PRESIDENT, S.M. SIMPSON LTD.)

"We have business dealings with some people that have been in effect for 30 years. There is still no contract and as long as the other person is alive and I am alive there will be no need for one. There was no turning around and saying, "Now will that be VISA or cash"? We all miss that but the bankers are too close to our pockets now."

– GERALD RABOCH, (CHIEF OPERATING OFFICER, RIVERSIDE FOREST PRODUCTS LTD.)

"I remember when handshake deals were very common. There are still a lot of things done by handshake and just by word between the parties. It might start with just a phone call or a direct conversation and now it is usually followed up with a letter or some kind of contract. It is still a very small fraternity in the forestry group and a lot of agreements are still by handshake."

– ABE FRIESEN, (PRESIDENT, POPE & TALBOT LTD.)

"When I first started in logging there were no written agreements. Whatever the deal was, you stuck to it. The contract loggers, contract sawmilling, we never had anything written down. You would talk about it and shake hands and that would be it."

– TARA GHOG, (OWNER, ASPEN PLANERS LTD.)

"When I came into the business we never had logging contracts. We would just agree on a price, shake hands and go to work. But the dollars got bigger and more important and people felt they had to have legal protection. Finally we got to the point where two sets of lawyers would look at a contract and the forester and the logger would stand off to one side and wait."

– FRED STINSON, (ILMA PRESIDENT, 1977-79)

The LONG family remained a force in Thompson Valley sawmilling for a number of decades. In 1947, the Long's Merritt sawmill was sold to a firm called NICHOLSON AND CATES – out of Burlington, Ontario. For nearly 20 years, Ken Long ran another mill operation (FADEAR CREEK LUMBER LTD.) up the North Thompson near Barriere. In 1963 he re-purchased the then-expanded Merritt mill from the same owners who had bought him out 16 years earlier.

During the 50s, Long was instrumental in pulling together a number of independent mill operators in the Thompson Basin with the objective of combining their waste fibre to make pulp chips. As this initiative became reality, many sawmill burners were phased out. Slabwood had become a resource, instead of a disposal problem.

Ken Long continued to expand his operations into the 1970s. He retired from the industry after selling out to CANADIAN FOREST PRODUCTS LTD. in 1977.

Merritt Forest Products Ltd.

Merritt Forest Products Ltd. (MFP) was another operation dating back to 1950. This was originally a planer operation that processed rough lumber purchased from other mills. Such lumber was not hard to find. Forty years ago, many of the hundreds of small mills in the interior did not have planers. Thus, custom planing became a form of value-added manufacturing.

Among the names associated with this company in its early days were the McDiarmid brothers, Ken Johnson and Reed Banford.

The company was among the first in the area to hire a professional forester – Dave McInnes, a native of the Kootenay region and a 1953 graduate of the UBC forestry school.

"By 1965 we were about the largest producer in the area. We had three sawmills at that time – one at Lumby, one at Merritt and one at Avola. And we were, you know, buying out small outfits when we could. I had hired our forester from MacMillan Bloedel. Just a young chap out of university called Dave McInnes, he has just retired as President and CEO of Weyerhaeuser Canada."

– IAN MCDIARMID (ILMA PRESIDENT 1964-1965)

MFP bought the CANFORD LUMBER operation west of Merritt in 1957 and became MERRITT CANFORD LUMBER LTD. This company subsequently picked up WADE LUMBER of Aspen Grove and DIAMOND MILLS of Avola, along with LUMBY PLANING MILLS. All finally merged to form MERRITT DIAMOND MILLS LTD. This operation was eventually sold to Weyerhaeuser. Forester Dave McInnes stayed on as woods manager and eventually became president of Weyerhaeuser Canada.

Tara Ghog and Aspen Planers Ltd.

Another long-time figure in the sawmilling industry of the Thompson Valley was Tara Ghog who emigrated from India to Canada in 1940, at the age of 17. After several years on the coast he found work in a portable sawmill near Kamloops, owned by NANJO SINGH. Ghog later got his own mill enterprise near Cache Creek. He sold rough lumber to GATEWAY LUMBER CO. LTD. at Ashcroft, where GEORGE FILLINGER was mill manager. A few years later, Fillinger, Tara Ghog and some others bought a sawmill and planer mill from WOODFORD AND SONS LTD. of Merritt.

The operation was carried out on the same site now occupied by ASPEN PLANERS LTD. Expansion continued throughout the 1960s and 70s, along with changes in shares and partners.

In 1977 Tara Ghog became sole owner of the company now known as Aspen Planers Ltd. His sons, Surinder and Paul now run the family business. Another name in the history of this company is "Slim" Sidhu who for many years worked as sales manager for the company.

Netherlands Overseas Mills Ltd.

Owner NICK VAN DRIMMELEN operated portable sawmills in the Merritt area in the early 1960s. Later, the firm expanded to Chase and Revelstoke and eventually to Prince George.

Portable Mills – Hundreds At One Time

During the 40s and into the mid-50s, literally hundreds of portable sawmills operated in the B.C. interior. The term "portable" could be a bit misleading, suggesting a kind of mill-on-wheels. Portable mills were actually just small sawmills that could be disassembled and moved in short order.

An early portable mill.

Most portable mills were set up on logging claims. Lumber was hauled out and slabs and sawdust were left in the bush. From a practical point of view, it made eminent sense to saw round logs in the bush, and to move lumber out instead of cumbersome logs. There was also advantage to not having to burn slabs and sawdust at a central millsite in town.

Chippers were costly, and were economically practical only when fed with a huge supply of raw material. Portable mills did not produce enough volume to make on-site chippers worthwhile. Scattered in hundreds of locations, some very remote, the portables were unable to share costs of a central chipper.

Economic conditions were beginning to move the industry toward larger, centralized mills. The social benefits of working at a central millsite in town were also an attraction for most millworkers. In the heyday of the portables many millworkers were required to live in camps – away from families in town.

Through the 50s it also became apparent that centralized, high volume mills could process wood with greater efficiency than could a network of smaller mills scattered throughout a forest region. "Economies of scale" became a factor when better roads and trucks made it less costly to haul logs longer distances. Another factor that favoured the trend toward centralization was the use of "chippable" sawmill waste as a fibre source for pulp mills. Once a waste resource, chips were to become the sole source of fibre for southern interior pulp mills.

By the early 1960s, the Forest Service actively encouraged operators to utilize more of the tree and to reduce waste, rather than sending massive amounts to a burner or building huge slab piles. Today, some mills generate their own electrical power by burning wood waste. Some are even able to sell surplus power to electrical utilities.

Pulp mills now supplied from waste wood

From the industry's start, only a small percentage of the pulp manufactured in the B.C. southern interior was made from round wood (logs). Good sawlogs produce a far higher value, and generate more jobs when cut into lumber rather than chipped for pulp.

In the past few decades, interior mills have converted higher proportions of sawmill waste (slabs, edgings, etc.) into chips. This "new" resource now comprises the raw material for southern interior pulp mills. Making use of more of the tree volume has been a trend throughout the industry since the 1950s.

All pulp mills in the southern interior are supplied with chips made from waste wood. The only place where logs are chipped is at Cache Creek, where a chipper processes hemlock logs which are too decayed to make lumber. The chips from this mill are sold in the U.S.

Kamloops Gets A Pulp Mill: Weyerhaeuser Canada Ltd.

The WEYERHAEUSER mill in Kamloops, and others, were set up to process the huge and previously unused volume of waste wood fibre created by the Interior sawmill industry. The events which led to the building of the Weyerhaeuser pulp mill at Kamloops were filled with complications, including extensive bargaining and negotiations.

As mentioned earlier, Ken Long was one of the prime movers of this initiative. In 1957 Long and HUGH HODGINS OF CROWN-ZELLERBACH shared thoughts on the use of waste fibre from Thompson Valley sawmills to feed a possible pulp mill in Kamloops. Their vision for an Interior pulp mill had to be tempered with patience. While they waited, Long managed to convince a number of interior mill operations to chip waste wood for shipment to a Crown Zellerbach pulp mill on the coast.

The Kamloops pulp mill concept steadily picked up steam, as many companies and individuals became involved at the study and planning stage. KAMLOOPS WOOD CONVERSIONS LTD. had the ball for a while but no mill resulted. Crown Zellerbach completed a pulp mill feasibility study, but decided not to go ahead with the project.

B.C.'s Forestry Minister, RAY WILLISTON, was anxious to see a pulp mill established in the interior.[1] He was convinced that it was possible to get more value (and more employment) out of the volume logged by converting sawmill residue to chips. At that time (mid-to-late 50s) pulp mills were located only on the B.C. coast. None of the coastal companies seemed interested in building a mill "upcountry".

Finally, a group of investors which included PHOS BESSETTE, ART HOLDING and KEN LONG, took a first step. They formed a corporation called KAMLOOPS PULP AND PAPER LTD. and were eventually granted a pulpwood harvesting license. When no Canadian company could be persuaded to join them to provide technical expertise and capital, Weyerhaeuser became involved at the eleventh hour.

Weyerhaeuser later expanded within the interior, to acquire, among other companies, Blue River Sawmills Ltd., Merritt Diamond Mills, B.C. Interior Sawmills Ltd., Kamloops Lumber Co. and Northwood Properties Ltd. of Penticton.

[1] Ray Williston's 16-year term as Forests Minister (1956-72) was notable as a time of enlightened and progressive forest management. Under his leadership, close utilization and sustained yield policies were instituted in B.C.

In 1964, Weyerhaeuser obtained 51 percent of Kamloops Pulp and Paper and proceeded to build the long-awaited mill which opened in 1965. It made, and continues to make exclusive use of sawmill residue, including sawdust, an achievement which proved the benefits of using more of the tree.

Weyerhaeuser proved that pulp mills could operate with sawmill residue, a resource previously regarded as worthless. Now when sawlogs are made into lumber, the pulp industry is supplied with the by-product of sawmilling. Wood chips later became an important revenue feature to sawmills. In times of depressed markets, chip income kept many interior mills afloat.

In 1971, all the remaining shares of Kamloops Pulp and Paper Company were purchased by WEYERHAEUSER and the company became known as WEYERHAEUSER CANADA. RALPH THOMAS, who had done Weyerhaeuser's feasibility study on the project, became the first CEO of Weyerhaeuser Canada. By 1972, production had grown to more than 1,000 tons per day.

Weyerhaeuser became a fully integrated company in the 1970s and 80s, with its own forestry department (including a tree nursery near Vernon) and engineering capability. The company now produces one fifth of the volume manufactured by ILMA member mills. Weyerhaeuser Canada has expanded into Saskatchewan and Alberta, and a Richmond, B.C. unit collects high grade office paper for de-inking and recycling.

Much of Weyerhaeuser's progress occurred during Dave McInnes' 17 years as President and CEO. After his retirement in 1993, McInnes was succeeded by GEORGE WEYERHAEUSER JR. as President and CEO.

STEVE SERY, Weyerhaeuser Vice President of Canadian Lumber Business, is a member of the ILMA Board of Directors.

Lief Norgaard and Ardew Wood Products Ltd.

LEIF NORGAARD, an emigrant from Denmark in 1951, worked in the construction business in Merritt for several years. He and his wife ELEANOR formed a partnership with JACK DEWOLF to buy T & G WOOD PRODUCTS LTD., a moulding plant. In 1966 they incorporated as Ardew Wood Products Ltd. At Jack Dewolf's retirement in 1971, Leif and Eleanor purchased his shares.

In the mid 1970s, Ardew became "a research lab" of sorts for Dr. Suezone Chow of Western Forest Products Laboratory at UBC. Dr. Chow, with Ardew's co-operation, is credited with the development of the "pre-heat glue curing system", a major step in the evolving technology of "finger jointing" lumber.

With diligence and commitment, Ardew pursued foreign markets for its specialty wood products and succeeded in establishing strong business relationships in Japan. Ardew remains a family-owned business today. ERIK NORGAARD has been president since his father's death in 1987. Erik is a member of the ILMA Board of Directors.

> *"I started working full time with my dad when I was 16 years old. I learned a lot from my father. It was good and it was necessary..."*
> *"We were almost the first ones to ship heavily to Japan from the interior of B.C. We started in 1978. In the downturn of the North American market a lot of people have turned to overseas markets. In the interior 10 years ago, there was hardly anybody doing overseas markets."*
>
> – ERIK NORGAARD,
> (PRES., ARDEW WOOD PRODUCTS LTD.)

Sandur Brothers and NMV Lumber Ltd.

NMV LUMBER LTD. is a family-owned remanufacturing plant built in Merritt in 1983. The Sandur brothers, Ravinder and Sukhdev, own and operate the plant which has drying and planing facilities. Ninety-five percent of NMV's manufactured products are shipped to export markets.

Kamloops Lumber Company and Harry Turner

The owner/operator of Kamloops Lumber Co. from 1930 until 1946, HARRY TURNER was also a founding member and the first Chairman of the ILMA.

In 1946 Turner sold KAMLOOPS LUMBER CO. to THE PAS LUMBER CO. THE PAS resold to REVELSTOKE SAWMILLS in 1948. Ken Long took over as manager and held that position for about 20 years. In 1970, Revelstoke Sawmills sold the mill to KAMLOOPS PULP AND PAPER. It operated for another year in Kamloops before moving to Vavenby.

Kamloops Lumber Company Ltd. 1955.

B.C. Interior Sawmills Ltd. and Phos Bessette

Phos Bessette got this mill started in 1943 at Mission Flats, west of Kamloops. A planer operation at Ashcroft was carried on by subsidiaries of the Kamloops company. But changes in wood utilization standards made it more viable to consolidate in Kamloops.

The Bessettes provide an excellent example of a traditional family sawmilling operation. BILL MCCOID and DON MARRIOTT – sons-in-law of Phos Bessette – were, respectively, manager and accountant at the Ashcroft mill. DON MUNRO was responsible for forestry and logging. All served on various committees for the ILMA over the years. Phos Bessette was one of three partners in Kamloops Wood Conversions Ltd., a forerunner of Kamloops Pulp and Paper Ltd.

Gateway Lumber Co. Ltd. – George Fillinger

This Ashcroft planer operation dated from the early 50s. Owner George Fillinger eventually sold his company to the Moar family. George then formed a partnership with Charlie Bennett and Tara Ghog to buy Woodford and Sons holdings in Merritt. Woodford led, in turn, to the formation of Aspen Planers Ltd.

Terrace Forest Products Ltd and Halston Lumber (Kamloops)

The opportunity to chip waste material as resource for a pulp mill, instead of sending it to a beehive burner, accelerated many changes in interior sawmilling. Terrace Forest Products and Haltson Lumber provide an example.

CHARLIE BENNETT, along with silent partner GERALD MAIN, were accustomed to purchasing lumber from a number of small local mills. The chipping opportunity created a need to consolidate. The consolidation was accomplished in rather unusual and colourful terms. Bennett and Main offered to sell their operation for "$1 million and a black Lincoln Continental". The offer was accepted by BALCO FOREST PRODUCTS LTD. of Kamloops. After establishing as one of the major integrated companies in the Kamloops region, Balco was sold in 1987 to TOLKO INDUSTRIES LTD.

Balfour-Guthrie Sawmill

This mill operated in Kamloops for a short time during the 50s. JIM MCKINNEY and GEORGE FILLINGER were mill managers and Reed Banford was woods manager.

Frolek Ranch and Sawmills Ltd.

This family business, near Kamloops, got its start in the late 40s. Various family members managed the operation through roughly 30 years of existence. The company went out of the sawmill business in 1973.

Pondosa Pine Lumber Company Ltd.
One of the very early ones

Pondosa Pine Lumber Co. Ltd., Monte Lake, B.C. 1974.

Another founding member of the ILMA, WILFRED HANBURY established Pondosa Pine in the 1920s, at the north end of Monte Lake. In 1955 the company withdrew from membership in the ILMA over a controversy involving a freight rate issue. CROWN ZELLERBACH purchased Pondosa in 1970, and later shut the mill down and used the site for log sorting.

H. Buff Lumber Co.
Another Monte Lake sawmill

H. Buff Lumber Co. – 1960's.
(Photo by Kermode Studio, Vernon, B.C.)

HENRY BUFF, with sons Rudolph and Peter, cut lumber in the Prince George and Fraser Valley areas before moving to Westwold in 1950. In 1952, they began to operate their own sawmill on the south end of Monte Lake. Upon Henry's retirement in the late 1970s, his sons took over. Shortly after their mill was destroyed by fire in 1980 they sold their interests to Crown Zellerbach.

Kam-Pac Forest Products Ltd. Brian Swanson and Probyn Lumber Export

This company, incorporated in 1987, started with a remanufacturing plant at Kamloops. The mill cuts domestic and off-shore specialties. The output of the plant comprises about 25 percent to domestic markets and 75 percent off-shore, mostly to Australia and the United Kingdom.

Bert Balison Family – Heffley Creek
Later Balco Forest Products

The Balison family bought a small mill at Heffley Lake, 40 kilometers north of Kamloops in 1946. Logging was carried out with horses. Only a few bare facts are available on the early days of this family operation that eventually grew into one of the biggest mills in the Kamloops area.

In 1949, the Balisons moved the mill closer to the wood supply. Two more portables were purchased and the company built a planer at Heffley Creek. Following the trend away from portable mills in the 50s, the Balisons built a central mill at Heffley Creek, converted to tractor skidding, and changed their company name to BALCO FOREST PRODUCTS LTD. Destroyed by fire in 1958, the Balco mill was quickly replaced with a more modern facility.

Enter the Quota System

During the late 1950s the Forest Service instituted the so-called "quota" system of timber supply. Unlike the area-based Tree Farm License, the quota system was a volume based arrangement. Under quotas, available timber was assigned to existing operators. The amount of quota allocated to a company was based on an average of their cut on Crown timber sales, for the three-year period – 1957 to 1959.

Although quota was not easy to obtain, BALCO managed to obtain salvage rights in 1959 to a large volume of fire-killed timber, at a place called Clemina, up the North Thompson River. Because they were willing to take the dead wood, the company was also successful in getting quota in green wood. For a few years the company operated a sawmill and planer at Clemina. When log hauling became more economical, the company closed this mill and moved the wood to their Heffley Creek location, first by rail and as roads improved, by truck.

During the 60s, Balco merged with TERRACE FOREST PRODUCTS AND HALSTON LUMBER and then went public in 1967. Following several major changes, the Balison family sold the company to Canadian Forest Products Ltd. (CANFOR) in 1977.

During this transition period, TREVOR JEANES was Balco's professional forester. In a cooperative arrangement with the Forest Service, Jeanes helped to establish the Balco Reforestation Centre. This centre now produces approximately 11 million seedlings per year. Balco Forest Products was sold to TOLKO INDUSTRIES LTD. in 1987.

J.N. Newberry and Sons

This family operation had a mill at Barriere Lakes from the 50s. In 1965 they sold out to Ken Long's Fadear Creek Lumber.

Darfield Building Products Inc.
Paul Shilling and Ray Inkster

Darfield is located midway between Barriere and Little Fort in the North Thompson valley. Started in 1972, two years later this mill was destroyed by fire, at one time a common occurrence in the interior dry belt. A year after the fire, GARY BURKHOLDER bought out RAY INKSTER and the mill was rebuilt. Burkholder and PAUL SHILLING remained as partners until 1991.

Yellowhead Remanufacturing
Gary Nelson

Located about 14 kilometers from Barriere on Highway #5, this remanufacturing operation came into being in September of 1992. Strictly a planer mill, Yellowhead custom planes for lumber brokers and other customers. The company puts rough lumber through a re-saw for remanufacturing into decking, channel siding, bevel siding, V-joint and dimension lumber.

Clearwater Timber Products Ltd., The Swanson family

Clearwater Timber Products Ltd., Clearwater, B.C.

The Swanson family has been involved in western Canada's forest industry for close to 70 years. Bob Swanson operated in Alberta until 1940, when he began to find it difficult to get timber in the province. He moved to B.C., formed Clearwater Timber and was successful at getting cutting rights in the Thompson Valley. Frank Capostinsky, a contract logger for the Swansons in Alberta, came with them to the North Thompson to oversee the logging side of the business.

Capostinsky started his own company – Thompson River Logging. He sold to Clearwater Timber, his one-time employers. After starting another sawmill at Blue River that went broke, he left the industry.

Fred McLellan, who started on the green chain at the Vavenby operation of Clearwater Timber, rose to the position of manager of the company operations at Clearwater, Vavenby and Valemount. McLellan was President of the ILMA in 1986 and 87.

The Swansons sold their operations to Slocan Forest Products in 1987.

Merritt Diamond Mills

Ian McDiarmid and Roy Gould

In 1950, Ian McDiarmid forged a deal with Brownsville Timber of New Westminster. He would cut up the log inventory stored at their then-idle mill at Avola on the North Thompson. In 1961, they merged the Avola mill with operations at Merritt and Lumby, calling it Merritt Diamond Mills Ltd. The holdings were sold to Kamloops Pulp and Paper in 1965. When Kamloops Pulp and Paper's new mill was built at Vavenby in 1971, the Avola plant was closed.

Stern Wheeler, 'Florence Carlin' booming logs in Shuswap Lake, circa 1900s. (Courtesy Salmon Arm Museum)

A 'tight' corner – Mark Dumont & Sons, Wilson Lake, 1957.

The Value of a Sense of Humour

Among working people everywhere, in virtually every occupation, the gift of humour is often called upon to keep up the spirits of co-workers. In everyday, or even dangerous aspects of their work, people of light-hearted disposition help us all to resist taking life too seriously. A sense of humour, even under the most trying of circumstances, has been a necessary tradition in the forestry business.

"One situation I remember that called for a sense of humour centred around the IWA strike of 1967 – the longest labour curtailment in ILMA record. Earlier that summer, the largest forest fire in the province had occurred within the operating area of Federated Co-operatives Limited (FCL). The fire had apparently resulted from a tourist campfire which quickly got out of control due to abnormally high winds during the summer period. The fire, called the "Mag" fire, ultimately burned 55,000 acres and covered three mountain ranges in one evening. FCL had the major part of the task of salvaging the timber that fall and winter, while the strike was on. As a result, all supervisors took to the bush, along with the office people and everyone else who was not in the union. We ran everything from tug boats to graders, did the log scaling, etc. There were quite a few instances of consternation with supervisors supervising supervisors, one of the most impossible situations imaginable."

– CHARLIE McCARTHY, (SENIOR VICE PRESIDENT, FEDERATED CO-OPERATIVES LIMITED)

"In 1937 I was about 13 years old and of course I thought I was a man. Some girls came up to the mill one week-end and I had on just my jeans and caulked boots and my long underwear and a belt – no shirt. I was kind of showing off and I

stepped over this spinning shaft and this set-screw hooked the cuff off my pants and just as fast as you could say "snap" it stripped me – yanked the pants right off me. There I was with a belt on, underwear and boots but no pants. And all those girls watching! I think there must have been two or three of them anyway."

– KEN HUSCROFT, (PRESIDENT, J.H. HUSCROFT LTD.)

"This is a story about the portable mill I started with. There was a drive shaft that drove the mill. I warned everyone to stay clear of it. One day when I was shoveling sawdust I tried to go under it while it was still turning. My shirt hooked on the drive shaft and it tore my shirt, pants and all my clothes off except for my boots, socks and shorts. I was lucky that I had changed into my light clothes that day. I ended up on a sawdust pile about 20 feet away. I was lucky I wasn't killed."

– TARA GHOG, (OWNER, ASPEN PLANERS LTD.)

"When I was a forestry student I worked for Gordon Burns, the general manager of Passmore Lumber in the Slocan Valley. Gordon said, "Take this crosscut saw and cut those windfalls so the 'Cat' can build the road." I got the saw stuck in the cut and asked the foreman what to do. He handed me a double-bitted axe and told me to chop the saw out. I went back to the log and started to do that and all of a sudden there was a "ping" and I had two pieces of saw. I didn't know if I was going to get fired or not but I took the two pieces of saw back to the foreman. Of course he went into a rage and said, "You're just a dumb university student. You fellows shouldn't be allowed in the woods."

– JOHN MURRAY, (RETIRED VICE PRESIDENT, WOODLANDS DIV. C.F.I.)

Dumont & Sons Early 1950s. "Oops!"

Crawford's Portable – 1906. Kelowna area. Photo by G.H.E. Hudson. (Courtesy Kelowna Museum)

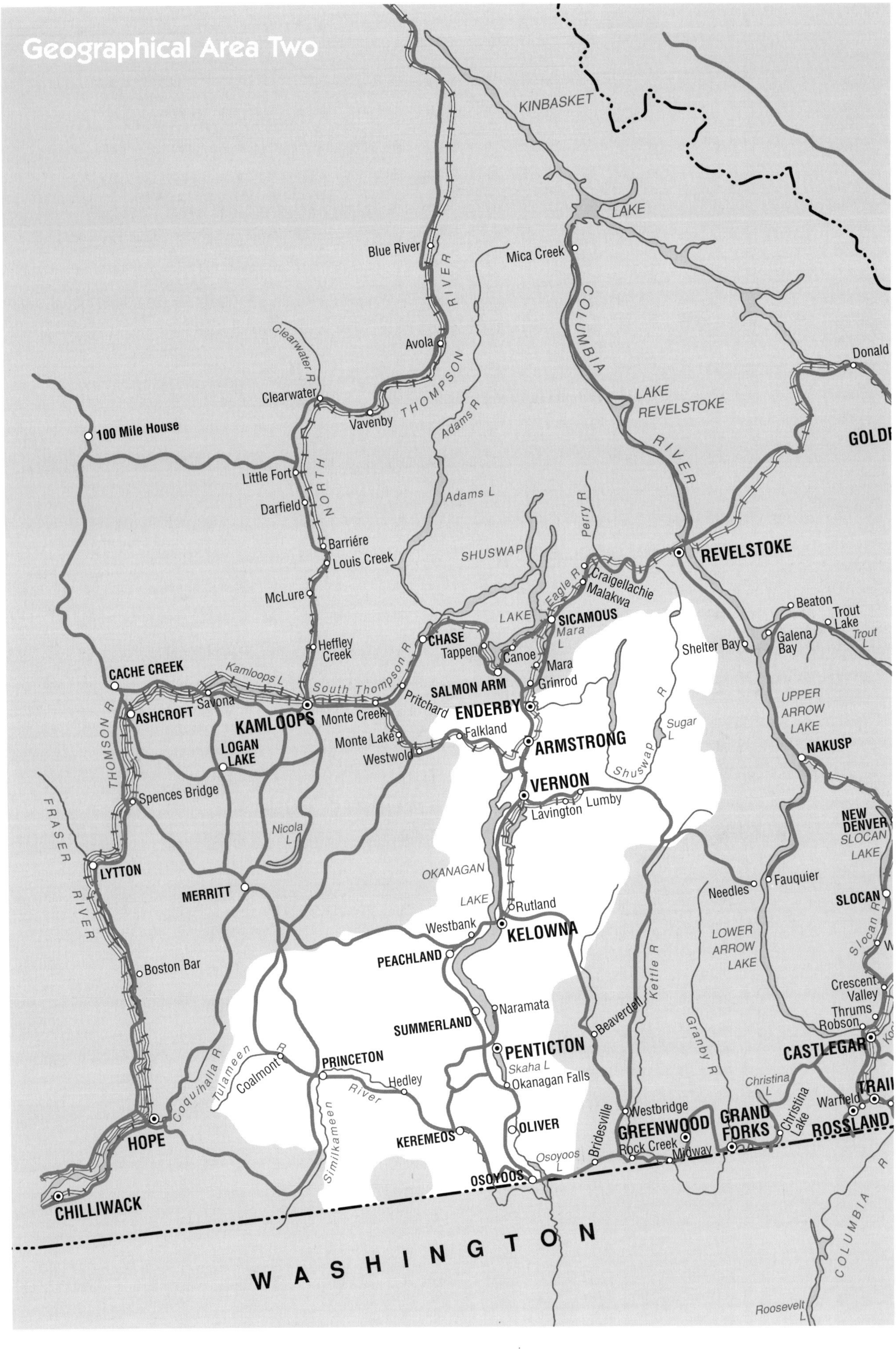
Geographical Area Two
KINBASKET
LAKE
Mica Creek
Blue River
COLUMBIA
RIVER
Avola
Clearwater R
Clearwater
Vavenby
NORTH THOMPSON
Adams R
Donald
LAKE REVELSTOKE
100 Mile House
Little Fort
Darfield
Adams L
Barriére
Louis Creek
SHUSWAP
Perry R
REVELSTOKE
Craigellachie
Malakwa
Eagle R
McLure
LAKE
SICAMOUS
Mara L
Beaton
Trout Lake
Galena Bay
Shelter Bay
Heffley Creek
CHASE
Tappen
Canoe
Mara
CACHE CREEK
Kamloops L
South Thompson R
SALMON ARM
Grinrod
UPPER ARROW LAKE
ASHCROFT
Savona
KAMLOOPS
Monte Creek
Pritchard
ENDERBY
Sugar L
LOGAN LAKE
Monte Lake
Westwold
Falkland
ARMSTRONG
NAKUSP
THOMSON R
Spences Bridge
Shuswap R
VERNON
Lavington
Lumby
FRASER RIVER
Nicola L
NEW DENVER
SLOCAN LAKE
LYTTON
MERRITT
OKANAGAN LAKE
Needles
Fauquier
Rutland
SLOCAN
Westbank
KELOWNA
LOWER ARROW LAKE
Slocan R
Boston Bar
PEACHLAND
Kettle R
Crescent Valley
Naramata
Thrums
Robson
SUMMERLAND
Beaverdell
Granby R
CASTLEGAR
Coquihalla R
Tulameen R
Coalmont
PRINCETON
PENTICTON
Skaha L
Christina L
Hedley
River
Okanagan Falls
Warfield
Similkameen
Westbridge
GRAND FORKS
Christina Lake
OLIVER
Bridesville
GREENWOOD
ROSSLAND
HOPE
KEREMEOS
Osoyoos L
Rock Creek
Midway
OSOYOOS
CHILLIWACK
WASHINGTON
COLUMBIA R
Roosevelt L

Geographical Area Two Okanagan/Similkameen

The Okanagan lumber industry started at about the turn of the century, following the bustle of railroading, and then supplying dimension lumber for local construction. Gradually an export lumber industry began to build.

The lumber industry in the Okanagan, and to a certain extent in the West Kootenay region, differed from other regions because of the burgeoning fruit industry. Until quite recently, almost all fruit products were shipped in wooden boxes. The growing demand for crates and bins created a reliable market for the products of Okanagan Valley sawmills. The sawn material used in making the fruit boxes was called "shook" – a word of unknown origin.

Following a similar pattern to the milling history of other regions, the Okanagan had many small and portable mills in the early years. These gradually became fewer in number as centralization occurred. Several family operations were eventually sold out to integrated companies, continuing a trend that was duplicated throughout the interior. The trend continues, on a reduced scale, today.

Penticton Sawmills and Hughie Leir

HUGHIE LEIR emigrated from England in 1902 and started a sawmilling operation near Penticton in 1905. The mill was moved to a new site on the Okanagan River south of town, where it remained for the next 50 years.

As did a number of other mills in the interior, Leir's Naramata operation moved the wood to booms by way of a creek-fed log flume. THE KETTLE VALLEY RAILROAD (built to connect the Okanagan and Kootenay regions) was completed in 1914 and changed the way many interior mills operated. Leir's mill was no exception. He could now ship logs to the mill by rail, from the Beaverdell area in the east, and from the Princeton area in the west.

Leir's operation became known in 1941 as PENTICTON SAWMILLS. The sawmill was powered by steam, generated by boilers fed with hog fuel (slabs, bark and sawdust). The mill cut lumber and produced box shook for the fruit industry.

From forestry's earliest days, a few operators purchased private forest land whenever possible. Many relied upon the security and long-term economic sense of ownership, as opposed to total dependence on the availability of crown timber through the Forest Service or the purchase of logs from private timber holdings. Apparently one entrepreneur who believed in private ownership, Leir eventually accumulated about 2,800 hectares (7,000 acres) of good timber-growing land that supplied his mill's timber requirements over many years.

A revealing anecdote from Leir's days shows how early members of the industry gained a reputation for being a feisty lot. Howard Jeal, who worked for Leir from 1928 to 1941, recalls a day when tempers flared around the millsite. A Scottish mill superintendent named Tom Rae punched Leir, knocking him over a load of lumber. It was the fight's one and only punch. It was not surprising that Leir responded by firing Rae. As it turned out, the careful Scot had secured another job before picking the fight. The biggest difference between such an incident occurring then and now is that apparently no police or courts became involved. The event was considered of concern only to the actual participants.

Loggers throughout the U.S. and Canada reputedly settled disputes by physical means. In the Maritime provinces, there was a reason for calling caulk boots 'Cape Breton boxing gloves'. (For some mysterious reason 'cork' is the common pronunciation for caulk.) Most fights took place in towns and usually followed the partaking of spirits. Normally these were not personal grudge brawls. The younger men of the industry were more likely to get into a fracas just to blow off steam.

Penticton Sawmill. 1928.

Hughie Leir's sons, John and Paul, took over Penticton Sawmills when he retired. The mill was sold to Northwood in 1966.

Loading logs at Mara. 1928.

Northwood and Noranda

Northwood Mills (headquartered in Prince George) became Northwood Pulp and Timber Limited in 1964. Two energetic personalities, Alf Powis and Adam Zimmerman built a giant forest-based corporation which at one time included controlling interest in half of Northwood Pulp and Timber, as well as in pulp and paper mills and sawmills in Quebec and New Brunswick.

Doug Little, who worked for the company and its forerunners for nearly three decades, was also a major figure in the building and ongoing management of Northwood.

Cooke Lumber of Greenwood, Western Pines Lumber Company of Princeton, Keremeos Sawmills and Abernathy Sawmills, both of Keremeos, Penticton Sawmills and B. Ellett Lumber Co. of Beaverdell all came under Northwood control in the 1960s.

Other names associated with Northwood or its predecessor companies were Ray Dewar, Henry Olson, Slim Perrault, and Fred Stinson.

During this rapid growth period, Northwood, along with all forest companies, was feeling the impact of new close utilization requirements. To meet the requirements, the company built two sawmills to handle small logs, one at Princeton and the other at Okanagan Falls. (Both were subsequently sold to Weyerhaeuser.)

Northwood and other companies continued to consolidate through the 70s. Changes in mill ownership and often in cutting rights occurred frequently as companies adjusted in order to stay competitive through rapidly changing times.

Clarke's Sawmill
William Albert Clarke

William Clarke built a sawmill in Penticton in 1935. At first steam-powered, it was later converted to diesel. The mill burned to the ground in 1939, and a year later Clarke built a planer mill which took rough lumber from bush mills at Yellow Lake and from Anarchist Mountain. In 1947/48 Clarke built a new sawmill on the Penticton Indian Reserve. He also ventured into the retail building supply business in Penticton. His mill operation remained in the family until son, Bud Clarke, sold out to Joe Dobie in 1967.

Yellow Lake Sawmills and Greenwood Forest Products Ltd.
Joe Dobie and Peter Buelah

Partners Joe Dobie and Peter Beulah ran the above sawmills during the 70s. It was during their time that the former Clarke planer was moved to the Penticton Indian Reserve site.

When Yellow Lake and Greenwood declared bankruptcy during the recession of the early 1980s, their quota was taken over by Gorman Bros. of Westbank. By 1983 the two long-time partners had recovered to open a remanufacturing plant on their original mill site. The plant produced panelling, furniture components and mouldings for markets in Canada, the U.S. and Britain. Peter Buelah has continued to operate the reman plant since the death of Joe Dobie in 1983.

Other South Okanagan Operations

Osoyoos Sawmills Ltd., was started by the Jorde Brothers in the mid 30s. Several partners took over in the early 40s. When the mill was destroyed by fire a few years later, the partners sold to National Forest Products Ltd., which also acquired Oliver Sawmills Ltd., along with some other mills in the area. In 1961 National sold to Northwood Mills Ltd., a subsidiary of Noranda Mines. This was the first appearance of Noranda in the forest industry of B.C. The corporation eventually became a major presence in the interior industry.

Names associated with those early operations in the south Okanagan include J.F. Nesdoly, Harold Wright, H.B. Collins and forester George Minns.

Oliver Sawmills. Late 60s.

Working with a "Cross-Cut" saw. (Courtesy Kelowna Museum)

Summerland Box Co. Ltd.
W.A. (Bud) Steuart

Summerland Box Co., operating in the 1940s, apparently started out by cutting material for the fruit box business, and then diversified into dimension lumber. The company was later acquired by Northwood, which used the log supply for its mill at Oliver.

Agur Logging Co. Ltd. and Tyac Box Co. Ltd.
Don and Pat Agur

A logger for SUMMERLAND BOX during the 50s, DON AGUR ended up with his own mill at Link Lake west of Summerland. In 1967, he sold the mill to STEELE LUMBER BROKERS. A year later he built a mill beside his brother's company in Summerland. PAT AGUR owned INDEL INDUSTRIES LTD., a contract sawmill that specialized in the box business. The brothers left the sawmill business in 1971. Don Agur has since died. Pat is living in Summerland.

Boys using peaveys to turn logs. (Courtesy Kelowna Museum)

Summerland Forest Products Ltd.

Summerland Forest Products was incorporated by John Gorman in 1982 to provide custom re-sawing, planing and drying services. Finger jointing, an important value added activity from the onset, has since been significantly expanded. The capability for edge glueing, another value added process, was added in the late 1980s. Presently this expanding enterprise is entering the laminated beam business in anticipation of further market opportunities in Japan. A related business at the mill site is a wood stove pellet plant, currently producing 30 tons of pellets daily from waste mill residue.

Other Small Regional Mills

Other operations of relatively short duration included NELS JOHNSON'S mill, located at "20-mile" west of Summerland, the T & V MATKOVITCH tie mill at Summerland, JACK HUGO'S mill, also in the Summerland area and PEACHLAND SAWMILL AND BOX CO. LTD, operated by NEIL WITT and TINY DELL. Further west was a mill owned by LEN TRAUTMAN and JACK GARROWAY. The latter operations were purchased by S.M. SIMPSON LTD. in 1955.

R. Nalos and Associates
Keith Marten

This mill had a short life span in the 1960s, at Faulder, west of Summerland. When the mill was destroyed by fire, owner Keith Marten did not rebuild.

Glenrosa area – site of present fire hall. 1908. Photo by A. Murchie. (Courtesy Kelowna Museum)

Gorman Bros. Lumber Ltd.
Ross and John Gorman

John (left) and Ross Gorman. 1951.

Born and raised at Glenrosa near Westbank, Ross and John Gorman entered the orchard business after serving in the Second World War. Their orchards were wiped out by the severe winter of 1949/50, and out of necessity, they entered the fruit box business.

> *"We started out making fruit boxes. In Westbank we were about the largest employer, and we only had about 15 working in all. Because we operated all year people wanted to work for us because the fruit industry was seasonal work. We have had a continuous operation since the beginning of 1951, except for shutting down for hunting season. In those early years, as soon as we got the first fall of snow we'd shut down and we'd all go hunting."*
>
> – Ross Gorman, (Managing Director, Gorman Bros. Lumber Ltd.)

Original Gorman Bros. Operation. 1952.

Initially the Gorman Bros. Box. Co. was strictly a backyard business. Ross and John bought trim ends from nearby mills and used a tractor power take-off for an energy source. By the late 50s they were sawing their own material and selling some lumber. They devised the "Gorman Bin" a practical collecting container for fruit pickers. Their bin could hold about 800 pounds of fruit, was transportable by fork lift, and is now in quite universal use in the orchard industry. In 1965 the company built a plant in Washington state to manufacture the bins.

> *"There were two sawmills in Peachland and ours in Westbank. There was also the bigger sawmill in Kelowna, S.M. Simpson Ltd. They bought the Peachland Sawmill and Trautman Garroway Sawmill. Maybe we weren't big enough at that time, because they did not try to buy us out."*
>
> – Ross Gorman

Also in 1965, the Gormans moved more intensively toward sawmilling, since it was obvious cardboard fruit boxes would eventually replace wood. They built a new sawmill in 1967 and a kiln in 1969, the same year they suffered a destructive fire. The fire provided an opportunity to rebuild and modernize. Through the 70s, Gorman Bros. placed its emphasis on producing one-inch lumber. The company now produces 72 million board feet per year. It also gets better utilization from wood by finger-jointing short trim-ends into longer boards – a technique which produces a high quality product from marginal raw material.

With the 1982 purchase of McGillis and Gibbs pole plant in Lumby, Gorman Bros. became cedar pole producers. The Ross Gorman family bought the interests of Skalbania and Nimilowicz in Downie Street Sawmills Ltd. (in Revelstoke) in 1990.

> *"We have been aware of the environmental issues, especially in the last 10 years. Our company seems to be in the middle of the controversy. We are logging in just about every watershed in the Okanagan. We have had to work long and hard to deal with environmental issues. A change for the better is everyone's respect for natural resources. It has been brought home to us. The industry was slow in learning and has paid a dear price for it."*
>
> – Ross Gorman

Gorman Bros. Lumber Ltd. has been an active supporter of the ILMA since 1961. Ross Gorman served as Chairman (1989 to 1991) and is a current member of the Association's Board of Directors. Over the years, a number of the next generation have progressed through the company to senior positions. The founding brothers continue to live in Westbank, not far from their Highway 97 millsite.

Horse logging at 8 Mile on Hwy. 33, Rutland, Winter, 1931. (Courtesy Kelowna Museum)

S.M. Simpson Ltd.
Stanley Merriam Simpson

This historic Kelowna-based operation got its start in 1913, the year S.M. SIMPSON left the construction business and bought a small sawmill. Just 27 years old when he founded S.M. Simpson Ltd., he cut fruit box material, continuing through the 1920s. In the early 30s, he expanded into the plywood and retail lumber business. Fire destroyed both the plywood and lumber plants in 1939, and both were immediately rebuilt.

Simpson acquired the Kelowna Sawmill Company about this time, but in 1943, it too burned. The Simpson Sawmill was expanded signficantly in 1950. The following year the company was awarded Tree Farm License No. 9, ensuring a reliable long-term wood supply.

In 1955, HORACE SIMPSON took over from his father. That same year he bought PEACHLAND SAWMILL AND BOX CO. from NEIL WITT, and then the TRAUTMAN AND GARROWAY operation at Peachland. In 1957 the company bought MCLEAN SAWMILLS in Malakwa and built a new plywood plant on the Kelowna site.

S. M. Simpson sawmill. 1940/41.

The Simpson enterprise grew steadily between 1957 and 1965. LUMBY TIMBER COMPANY was purchased and one of the first "chip-n-saw" machines in the province was installed at the Kelowna millsite. Simpson chips were shipped by rail to Crown-Zellerbach's pulp mill on the coast.

Stanley Merriam Simpson, a founding member of the ILMA in 1941, and Chairman from 1949-1951, remained a strong supporter of the association, along with many of his employees, until his death in 1959 while on a trip to Australia. His supportive tradition continued after Horace Simpson assumed management of the company.

In 1965, Crown-Zellerbach Canada Ltd. bought out S.M. Simpson Ltd. This was the first of many such major acquisitions in the southern interior. Almost 30 years later, through a complicated series of transactions involving Fletcher Challenge and others, the original Simpson operations would become part of Riverside Forest Products Ltd.

Crown-Zellerbach Canada Limited

The S.M. SIMPSON holdings in the Okanagan were the first interior B.C. operations acquired by CROWN-ZELLERBACH. The company ventured further into the Okanagan in 1969, purchasing ARMSTRONG SAWMILLS operations at Armstrong, Enderby, Vernon and Falkland. Another mill, PONDOSA PINE LUMBER CO., was purchased in 1970. Crown-Zellerbach closed the mill and converted the site, at the North end of Monte Lake, to a log sort yard.

In 1972, Crown Zellerbach struck a three-member task force to look at timber supply in its operating area. Task force members JIM CHARPENTIER (ILMA President 1968-1969), BILL MCGHEE and ERNIE LILLOS concluded that a full-scale modernization program was needed. Their recommendation led to the construction of a new sawmill at Larkin near Armstrong and the up-grading of the plywood plant. Following the centralization trend that was affecting all western companies, Crown-Zellerbach installed state-of-the-art facilities in Kelowna. As soon as the new Kelowna equipment was running smoothly, the company shut down its plants in Peachland, Monte Lake, Falkland, Armstrong and Enderby.

Crown-Zellerbach became CROWN FOREST INDUSTRIES LTD in 1983. Later, Crown Forest became part of the FLETCHER CHALLENGE collection of companies. Fletcher closed the Lumby mill in 1989. In 1993 Fletcher Challenge sold its operations in Kelowna and Larkin (near Armstrong) to RIVERSIDE FOREST PRODUCTS LTD.

H.P. Mills Kelowna Ltd.
Hans Peterson

Hans Peterson is an example of a small operator who bucked the trend to sell out to larger corporations. Peterson chose to remain independent. Although he died in 1991, his original mill remains in the Peterson family and is currently managed by his granddaughter, KATHY KEARNE. The secret of H.P. Mills' continued presence through circumstances that closed the doors of many small mills, may lie in the family's ability to locate and secure market niches still economical for small mills. Still small by modern standards, H.P. Mills continues to produce dimension lumber as well as railroad ties and precision, end-trimmed studs.

Rutland Sawmills Ltd.

As were so many other mills in the Okanagan, RUTLAND SAWMILLS was built to cut material for wooden fruit boxes. Several fruit packing houses were supplied with boxes by Rutland Sawmills in the 30s and 40s. SLIM PERRAULT, general manager during the 50s, had been an employee of THE PAS LUMBER COMPANY in Manitoba. Perrault had come west when The Pas began to establish a B.C. presence.

In 1962 Rutland Sawmills was bought by JOHN AND MARY OLINGER, owners of Olinger Lumber at Carmi. Slim Perrault left their company to join Northwood as general manager. AL MCMATH became manager of the Rutland mill, where he stayed until the mill was sold to BOUNDARY SAWMILLS. At that time he left to become mill manager at KOOTENAY FOREST PRODUCTS in Nelson.

With changes in the industry, small local mills grew less able to compete and were gradually phased out of existence. Rutland Sawmills shut down permanently in the late 60s.

Louie Russo Sawmills – Kelowna

Louie Russo Sawmills provides a classic chapter in the history of family-operated sawmills in the interior. The chapter opened with JOHN RUSSO's contract with the CPR, to supply hand hewn ties for rail construction at Golden, B.C. With experience behind him, John moved to the Okanagan in 1946 and carried on supplying ties, this time to the CPR's Kettle Valley branch line until 1953.

The Russo family also began to acquire orchard property in the Okanagan. In 1956, son LOUIE RUSSO ventured into the sawmilling business. He started small, setting up a portable mill on orchard property on the east side of Kelowna. Louie bought his father out in 1960 and started a planer operation, but continued to renew tie contracts with CPR.

The Russo family's 80 acres of orchard holdings, and the sawmill, are still run by members of this enterprising family.

Vernon Box and Pine Lumber Co. Ltd.

Dating back to the Great Depression decade, VERNON BOX AND PINE LUMBER CO. was a pioneer sawmill in the Okanagan region. JIM STROTHER, a founding member of the ILMA, was associated with this mill's earliest days. Located on Highway 6, the company continued operations until 1962 when it was purchased by ARMSTRONG SAWMILL LTD. Armstrong sold to Crown-Zellerbach during that company's expansion in the region in the late 60s and 70s.

ILMA founding member Jim Strother (seated on spruce log).

Riverside Forest Products Ltd.
Jerry Raboch, (son) Gerald and Gordon Steele.

The Raboch business story began in 1947, with the set-up of a portable mill. The family concentrated its horse logging operations in the Trinity Valley, east of Enderby. Over the next two decades, from these modest beginnings, the Raboch business would evolve into an impressive collection of companies and grow to include several of the best known southern interior mills.

From left Henry Raboch, Gerald Raboch, Alvin Raboch, Jerry Raboch (Photo courtesy Enderby and District Museum)

Riverside Forest Products Ltd., Ashton Creek site. 1963.

The fledgling Raboch company logged cedar poles that fetched high prices. Power authorities needed thousands of poles to deliver electrical service across the interior's vast rural landscape. As the company grew in strength, it was able to buy out various area mills. It was also astute in developing partnership arrangements with many well-established figures in the sawmilling industry.

From left Roy Boyd, Jerry Raboch, Joe Ornst, 1926.

At JERRY RABOCH'S retirement in 1959, his son, GERALD, took over the business. At the same time, the lumber business throughout the region began to change rapidly. In 1962, Gerald entered into partnership with WILLIAM H. STEELE LUMBER COMPANY of Vancouver. Their new company, RIVERSIDE FOREST PRODUCTS LTD., brought together the production and marketing aspects of the business. BILL STEELE and Gerald shared the optimistic belief that opportunities for expansion still existed. They agreed that they needed a larger asset base to ensure future growth. Riverside Forest

Products expanded to plants in Vernon and Lumby, with acquisition of Monashee Lumber Company, Fisher Planing Mills, Long Lake Lumber and Balestra and Schnyder. Bill's son, Gordon Steele, became active in the firm in 1965 and later took over the presidency of the thriving company.

In late 1992 the company purchased the southern interior assets of Fletcher Challenge – including sawmills and plywood plants, a nursery and seed orchard, and a wood container plant at Kelowna. Today, with Gordon Steele and Gerald Raboch at the helm of the management team, Riverside Forest Products employs more than 1,100 people in five locations. The current Chairman of the ILMA, Morris Douglas, is Riverside's Vice-President of Sales and Marketing.

"When we needed a new employee someone would just bring along a son, cousin or brother. Riverside was an extension of a large family group. We have just completed the family tree of Riverside Forest Products. Since 1955 we have combined with 35 different companies. Most of the elder statesmen of these companies wanted to retire and there was no one who wanted to continue on."

– Gordon Steele. (President & CEO, Riverside Forest Products Ltd.)

Gordon Steele was one of the founding directors of Fibreco Export, a consortium of interior companies established to export chips overseas. Gerald is President of Cantree Plywood Corp. in Vancouver. Cantree is jointly owned by Riverside and Slocan Group. Still an active member of the ILMA, Gerald served as its President from 1981 to 1983. He has had a Riverside "family tree" developed to chronicle three decades of growth.

Riverside has announced plans to purchase Jacobson Brothers Forest Products Ltd. The Williams Lake company, employing 300 people, produces random length lumber and has extensive timber holdings in the Cariboo region of B.C.

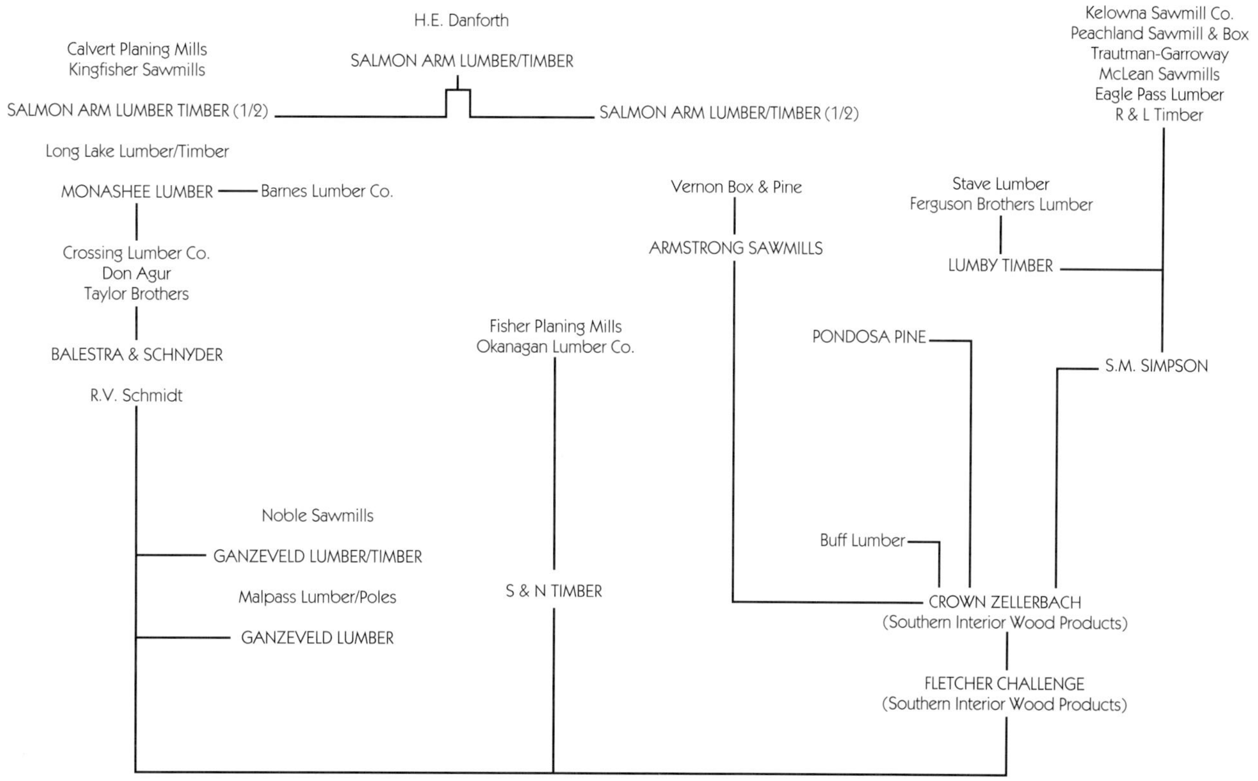

RIVERSIDE FOREST PRODUCTS – HISTORY

The Industry Continues To Change

Beyond statistical data and information – mills opening and closing, partnerships forming and dissolving, companies centralizing, modernizing and upsizing – the development of interior communities provides another historical record. In the 25 years following the Second World War, just about every Interior community had at least one sawmill. Owners of these sawmills were generally self-reliant, some possessing amazing mechanical and technical skills. It gradually became apparent, however, that those skills would no longer be enough. As competition grew more intense, many smaller mills found it difficult, and then virtually impossible, to survive.

To a large extent, markets were the impetus for change. Many portable mills could not cut lumber to precision standards, or even to uniform sizes. The changing market demanded ever higher standards of quality. It had become a buyer's market and buyers wanted high quality and consistent standards. The ability to meet lumber grading standards meant the very survival of the interior industry and was a major reason for the formation of the ILMA. It was a matter of high priority for the new association to agree on lumber grading standards, to ensure that member mills met the standards consistently, and to develop a reputation for delivering quality products.

Other factors transformed the industry during this period and gradually led to the disappearance of the small mill era and a way-of-life. Changing demographics had an impact. By the late 1960s, many mill owners, often World War veterans, were nearing retirement. Similar to the trend on family farms and in other small businesses, sons and daughters of sawmill owners were often not willing to take over the arduous and uncertain businesses that had caused struggles for their parents.

At the same time, forests themselves had changed. The handiest timber claims had been logged first and it was necessary to reach farther into the stands for timber. Longer log hauls in the interior's varied terrain created massive road-building costs, often far beyond the financial means of small and medium-sized firms.

Another major change was a new forest policy that required companies to make use of more of the resource. Forest Service studies had revealed excessive waste in wood volume at several stages between the standing tree and the finished piece of lumber. The sophisticated technology that would utilize more tree volume was unaffordable for small operators. And in certain cases, the small-scale technology they needed was not even available – at any price.

Events occurring in the forest industry in the southern interior were merely a reflection of the enormous post-war transformation of North American society and economy. Social changes were painful for many, but there seemed to be no way to shift course. Over a period of about 15 years, beginning in the late 1950s, the cumulative impact of changes affected virtually every industry, throughout the continent. Family farms and small businesses, including hundreds of small sawmills, simply disappeared.

Unquestionably, many small sawmill operators would have preferred to stay in business. The period of upsizing displaced families and workers, and created instability in many communities. But there had been downsides, as well, to the small mill period. Jobs in small mills were seldom secure. Layoffs were subject to factors such as lost markets or bad weather. In a mill with a handful of employees (a crew of five was average for portables), the breakdown of a single piece of equipment could send the whole crew home. And throughout the era of small sawmills, the accident rate was high.

While some small mill operators were reluctant to sell, many were willing and motivated. Some were eager to put cash in their pockets as a return for their investment and labour. Others were ready to leave management and marketing problems in some other hands. Any of their employees fortunate enough to find work in the larger, more centralized mills, appreciated the shorter hours and better pay and benefits, not to mention the opportunity to live in towns with schools and other amenities at hand. The greater security of employment gave many workers a peace of mind they might not have experienced in the era of the smaller mills.

"A lot of places had quite large sawmills at one time. I am thinking about places like Lumberton (near Cranbrook), Wardner, Monte Lake, and places like Vernon, Nelson, Chase, Enderby, Trail, Oliver, Penticton. They all disappeared for one reason or another."

– HORACE SIMPSON, (RETIRED PRESIDENT, S.M. SIMPSON LTD.)

"I've seen a lot of mills go under or sell out. If you didn't change you didn't last."

– Bill "Willie" Schumaker, (Retired President, IWA, Local 1-423)

"What usually happened in the interior was a small company would buy out another small company. And then that company would pick up another small company and so on. An example would be the Merritt Diamond experience. And we have the Crestbrook Forest Industries experience, which came from several small mills just getting bigger and coming into one. There was Federated Co-op at Canoe, which absorbed several small companies and grew to a larger corporation. S. M. Simpson out of Kelowna was another example. In the interior of B.C. I know of only a few instances in which large companies came in and took over small companies. Usually, they were one-owner companies and when the owner had to quit because of taxes or other reasons then usually another small company would take them over."

– Gerald Raboch, (Chief Operating Officer, Riverside Forest Products Ltd.)

As centralization continued, partnerships were formed and mill ownership changed frequently. Versatile and adaptable operators rolled with the changes and accommodated to the new era. Still, many businesses failed despite consolidation and modernization.

To survive, sawmill owners and logging contractors also had to adapt to changing B.C. forest policy. Loggers were no longer just loggers. The Forest Service set a new course – giving industry greater responsibility for more stages of forest management, and recently adding responsibility for tree planting. In some states and provinces, forest service staff continue to build roads and plant trees after harvesting.

The B.C. Forest Service deemed activities such as road-building and forest renewal the responsibility of major forest companies. Skills required to grow new forests were vastly different from logging and processing timber. It became increasingly difficult for many local or family operations to maintain their operating niche in this fast-changing environment.

Most industry consolidations occurred before the mid-70s. Consolidations then slowed and the industry entered a more stable period. The transition itself and the speed with which it was allowed to occur are still matters of considerable debate. Certainly, without the internal and external pressures that forced modernization, the interior forest industry would not have kept a competitive edge in continental and global markets. In the long run, despite social dislocation and economic disruption, the transition resulted in greater job stability and better working conditions. Many forest workers are retired or nearing retirement after 20, 30 and even 40 years with the same company. Such long-term tenure was unheard of in the days of the bush portables.

Many markets established during this period continue to serve the industry well today. The interior industry can look back at the 60s and 70s as its heady decades of expansion.

Tolko Industries Ltd.

This well-established company emerged out of the Lavington Planer Mill operation, east of Vernon, in which Harold Thorlakson was a major figure. Sons Allan, John and Doug continued to expand the company in the 70s, and took over Ernst Sawmills of Quesnel in 1981. In 1987, Tolko acquired controlling interest in Balco Industries of Kamloops. In addition to its interior operations, Tolko Industries has holdings in North Vancouver.

Curly Noble, Dave Simard and Dal Lumber Ltd.

Curly Noble, a skillful planerman who produced first class lumber, started his own planer operation near Vernon in the 1940s. In the mid-50s, he sold his operation to Rex Ganzeveld and went into the resort business at Mabel Lake. Two years later, when the lure of sawmilling was too hard to resist, Noble and a partner, Dave Simard formed Dal Lumber Ltd. They bought an old Yates planer and began processing lumber from Simard's small sawmill at Mabel Lake.

The planer was strategically situated on the edge of the airstrip at Mabel Lake. Dave Simard recalls that small aircraft sometimes had to taxi around air-drying lumber piles to prepare for take-off.

Curly Noble bought a sawmill at Tappen, near Salmon Arm, in 1960 which he operated for a year or two. He died in 1963.

Early log hauling near Lumby. 1930's. (Photo courtesy Enderby and District Museum)

Early logging in Mabel Lake. Fallers (on springboards) are Bill Johnston (left) and Eli Johnston. Mr. Fraser looks on. 1898.

H. Sigalet and Company Ltd.

The history of H. SIGALET AND COMPANY involved many partners and different milling and logging enterprises – all springing from the energy and resourcefulness of HENRY SIGALET. A charter member of the ILMA, Sigalet started his forest industry business in 1922, with a pole operation at Sugar Lake. He remained in the pole business, in various capacities, for most of his working life. He started sawmilling in Squaw Valley near Lumby in 1934, and in 1940 formed H. SIGALET & CO. LTD. with a sawmill at Lumby. When the Squaw Valley mill shut down in 1946, many of his employees moved to the Lumby operation.

Along with FRED KIRKLAND and OSCAR NEIMI, Henry formed LUMBY TIMBER CO. in 1947. The next year, a number of smaller mills owned by these partners shut down and operations consolidated under Lumby Timber.

Various partnership changes occurred over the next few years until the operation was purchased by S.M. Simpson Ltd. (of Kelowna) in 1959.

Sigalet died in Vernon in December of 1972.

Bell Pole Co. Ltd.

Three generations of the Bell family have contributed to the success of BELL POLE, a unique forest company. M.J. BELL SR. started the family's first forest-related work in 1909. His grandson, MERT BELL III, is currently Chairman of the Board of the company, which originated in Northern Wisconsin.

The firm established a pole production plant in the Lumby area in 1925. In the early 30s, it began treating the butt end of poles to preserve against rot. Facilities to treat the entire pole were added later. The treatment process helps extend the useful life of poles. Natural oils in western red cedar make the species very long-lasting, and even without treatment, cedar poles and fence posts can remain sound for up to 50 years.

The largest producer of cedar utility poles in North America, Bell Pole is also a major producer of lodgepole pine poles.

Bell Pole's allowable annual cut of approximately 174 thousand cubic meters is harvested from eight timber supply areas. The elevation, soils and climate in several regions of B.C. are ideal for cedar stand management.

From its original plant in New Brighton, Minnesota, Bell has expanded to a plant in Carseland, Alberta, as well as to pole yards in Lumby, Haney, Revelstoke, Terrace and Conway, Washington. The current President Cyril Hopkins and General Manager Ken Stewart, have been solid supporters of the ILMA. Stewart has been an ILMA Director for a number of years and is currently Vice-Chairman of the Board.

McDiarmid Brothers, Ken Johnson and Merritt Diamond Mills Ltd.

Three men, all World War Two veterans, began their long-term association with a tie mill in Port Moody. Their venture evolved into several milling operations near Hope. One of their mills, in the Skagit Valley, processed wood cleared from the U.S. side of the Ross Dam reservoir. [The dam is owned by Seattle City Light Co. Proposals in the 1960s to raise the Ross Dam would have caused still further flooding on the Canadian side of the valley bottom. A huge groundswell of public opposition marked one of the first major environmental campaigns in B.C.]

The McDiarmid brothers and Ken Johnson moved to Lumby in 1950 to launch Merritt Diamond Mills Ltd. They expanded through the first half of the 60s until selling to Kamloops Pulp and Paper Co. in 1965.

Harvey McDiarmid was President of the ILMA from 1953-1955. His brother Ian held the same position from 1963-1965.

Pioneer Sash and Door Co. Ltd.

In 1942, Frank Becker, later to be elected as Mayor of Vernon, and Stuart Oldham bought the S.C. Smith Lumber Co. in Vernon. They renamed their company Pioneer Sash and Door. They continued to operate until 1975 when they sold to David Howrie Ltd. The plant closed in 1982 and the City of Vernon purchased the site.

The Ganzeveld Brothers and the Salmon Arm Planing Mill

Rex Ganzeveld, an early operator in the southern interior, opened a small sawmill in Vernon in 1946. He and his brother, Engel, sold out to Riverside Forest Products in 1965. The same year they formed a new company and took over Salmon Arm Planing Mills Ltd. In 1968 they sold the planer side of the operation to Dennis Marsh and Bill Pogomiller. The same year it was purchased by Tappen Valley Lumber, and then by Lignum Ltd. of Williams Lake. In 1970, Marsh and Pogomiller bought Malpass Lumber and Pole Ltd. This operation was bought by Riverside Forest Products Ltd. in 1979.

Ganzeveld mill crew on fork lift. 1952.
(Photo courtesy Enderby and District Museum)

Hoover Family Operation – Newport Creek

Art Hoover's father joined the business in the early 1940s, shortly after Art opened a mill on the Indian Reserve at Newport Creek, three miles from the north end of Okanagan Lake. Tie-cutting operators at first, father and son expanded into full production, with dry kilns and planer.

Hoover Sawmill – Newport Creek. 1968.

In 1953 Art's brother, Herb Hoover bought his share in the company. Two years later, another brother, Len, joined the company. In 1964, Hoover Sawmill Ltd. was bought by Paul and Marc Dumont, and in 1972, by Lavington Planer Mill Ltd. Shortly after, the mill was closed and the log supply transferred to Lavington's operation near Vernon. Herb and Len Hoover retired in Vernon.

Postill Brothers; and C. Brewer The Century Before

With origins in 1882, the Postill Brothers sawmill at Deep Creek near Armstrong is believed to be one of the first to be established in the Okanagan region.

Another pioneer operation, dating back to 1891, was established in Armstrong by C. Brewer. Brewer's mill was later sold to Wood and Cargill Company. It was then acquired by Tom Knight Smith and brothers George and Billy in 1901. This purchase established a family business which continued for 68 years, most of those years as Armstrong Sawmills Ltd. Billy Smith's sons, Jack and Laurie, carried out various ventures in the sawmilling business. Jack moved to Enderby in 1944 to run a mill purchased by Armstrong Sawmills. He continued to manage, modernize and expand the Enderby operation over the next 22 years. Jack became company president after his father's retirement in 1965. Laurie took over as manager of the Armstrong mill that same year.

The company purchased Vernon Box and Pine Lumber Co. Ltd, in 1962. It operated as an affiliate company until all the Smith holdings were sold to Crown-Zellerbach in 1969.

Tom Malpass – Teamster Turned Trucker and Sawmiller

Born and raised in the Okanagan, Tom Malpass hauled logs by team and sleigh to the Smith's mill in Armstrong. He began trucking logs in 1936. A service station operation was added to his assets in 1941. He continued in the trucking business, contracting the hauling of logs to Armstrong Sawmills and poles to both B.J. Carney Company and Bell Pole.

Malpass bought a sawmill in Ashton Creek near Enderby in 1946 and the next year formed a company called Malpass Lumber and Poles Ltd. In 1948 he bought property north of Enderby – site of a former brick yard. He built a planer and finished lumber from the portable mill at Ashton Creek, and built a new sawmill at the planer site in Enderby.

Malpass Crew, 1948. (Photo courtesy Enderby and District Museum)

Malpass built the first chipping plant in the Okanagan and sent chips by rail to the pulp mill in Kamloops. He sold his operation to Ganzeveld interests in 1970, and it in turn sold to Riverside Forest Products Ltd. Riverside closed the mill in 1981.

B.J. Carney & Company, Limited, Pole-makers

Poles being loaded on flatcars at Mara, 1931. B.J. Carney & Co. Ltd. (Photo courtesy Enderby and District Museum)

This pioneer company opened for business in 1915, in Sicamous, in the heart of the interior wet belt region. Lower elevations are ideally suited for growing cedar poles. Carney expanded with an office in Enderby in the mid-20s and bought poles from throughout the north Okanagan and Shuswap areas.

In 1949, B.J. Carney bought R.W. Bruhn's pole interests and extended further, opening poleyards and concentrating business in places such as Sicamous, Chase, Enderby and Nakusp. In the early days, Percy Farmer was the general manager, succeeded by Bud Shantz and later, Jack Herman.

The business re-organized as a limited company in 1952. The main offices of B.J. Carney Co. were moved from Enderby to Sicamous in 1989. The company currently operates pole yards at Sicamous and Nakusp, as well as in Idaho and Montana. An affiliated company, Canada Cedar Pole Preservers Ltd., operates a pole yard and treatment facility at Galloway.

Pole-makers – A different set of skills

The first wave of forest harvesting in the interior responded to the demand for railroad ties and lumber for mining towns. The second wave came with the advent of electricity. Tens of thousands of cedar poles were suddenly needed to hang wires that would bring electric power to towns and cities.

Pole logging required skills markedly different from sawlogging. Poles had to be hauled out full length – quite a challenge for teams of horses or early tractors. Because poles were commonly peeled of bark in the bush, May and June were ideal months for pole logging. Loosened by the upwelling of spring sap, bark slipped easily from the trees. Mechanical peelers were uncommon in the southern interior until the late 1960s. After trunks were peeled, the exposed sapwood was as slippery as axle grease, and poles were easier to move.

Cedar poles are easily broken and have to be felled and yarded with care. The longest poles are the most valuable, so veteran pole-makers take the time to be gentle. Loggers experienced at moving sawlogs, which are hard to damage by comparison, usually have to undergo a period of adjustment when they start yarding more delicate cedar poles.

Truckers also face a considerable challenge moving poles on highway hauls. Seventy, 80 and 90-foot lengths, still being harvested and hauled in the southern interior, require log trucks with extra long reaches. Truckers must be highly skilled to safely move such large, breakable high-value products.

Steel poles and towers are now used on main transmission lines. But on the thousands of miles of smaller urban and rural lines, treated wood poles have endured as the product of choice.

Rollie and Bud Hill Another Father/Son Operation

This father-son team started a small mill operation in Enderby in 1945, cutting railroad ties exclusively for the first few years of operation. In the late 50's, they concentrated on dimension lumber, sending most of their product to local markets in the Enderby area. They sold to Riverside in the mid-60s.

Huff Brothers Sawmills Ltd.
Princeton, B.C.

Brothers CHARLIE, HENRY AND ROY HUFF added mixed ranching, farming, and mining to the sawmill business they started in 1942. They sold their main sawmilling interests to Northwood in the late 60s, but continue to saw timber from private forestland.

Kettle Valley Lumber Co. Ltd.
The Munsie Family

JACK MUNSIE, who served as ILMA President in 1962-63, opened a mill in 1947, at Summers Creek, five miles north of Princeton. Munsie operated KETTLE VALLEY LUMBER CO. until 1966 when it was sold to NICOLA VALLEY SAWMILLS LTD.

An active supporter of the Canadian Forestry Association, Munsie died in 1988. His wife DOROTHY resides in Princeton and continues to attend the ILMA's annual meetings.

Kettle Valley Lumber Co. sawmill. 1950's.

Taylor Lumber Co. Ltd.
Jack Taylor and Sons

TAYLOR LUMBER CO. LTD. was incorporated in 1938 after JACK TAYLOR and sons GRAHAM, NORMAN, BERT and BOB took over a bankrupt sawmill on the ABE WILLIS property in Wolfe Creek, east of Princeton. They moved this mill to a site near the present Princeton golf course, where they continued to operate for 28 years. The first mill was steam-powered but later converted to diesel and then electricity.

After Jack died in 1961, his sons operated the mill for five years before selling to NICOLA VALLEY SAWMILLS LTD.

W.T. Squelch and Son Ltd.
Tulameen

W.T. SQUELCH opened this mill in 1947. His son, GRANT, had worked for Fraser Mills at Mallairdville before relocating in the interior. Father and son added a planer to their operation in 1954. In the late 50s, they sold to OLIVER SAWMILLS who closed the mill in 1960.

Western Pines Lumber Co. Ltd.
John Nylund

JOHN NYLUND was a sawmiller who arrived in Princeton country from Grant's Pass, Oregon in the 50s. He ran a planer mill located near Princeton, supplied by portable mills at Glimpse Lake, near Merritt, and Beacon Flats in Princeton. The portables were shut down in 1956. Nylund built his own mill in east Princeton, on the site of the existing WEYERHAEUSER CANADA LTD. mill.

Nylund sold to NORTHWOOD MILLS in 1963. The mill operated under its original name, WESTERN PINES LUMBER, until Northwood amalgamated its area operations in 1967. Northwood sold the mill to WEYERHAEUSER CANADA in 1978.

McInroy Sawmills
Another Family Operation

Located seven miles east of Princeton, on the north side of the Similkameen River, MCINROY SAWMILLS was owned by HOWARD MCINROY and sons CLARENCE and GAVIN. The company was an ILMA member until sold to O'NEIL AND DEVINE LTD., of Merritt. Howard and Clarence McInroy are deceased. Gavin still operates the family sawmill in its original location.

The Maud-Allen at Oyama, 1910. This famous little steam boat is shown booming logs on Wood Lake. The logs were then towed through the canal and hauled to Carswell's mill on Kalamalka Lake near Vernon. (Courtesy Kelowna Museum)

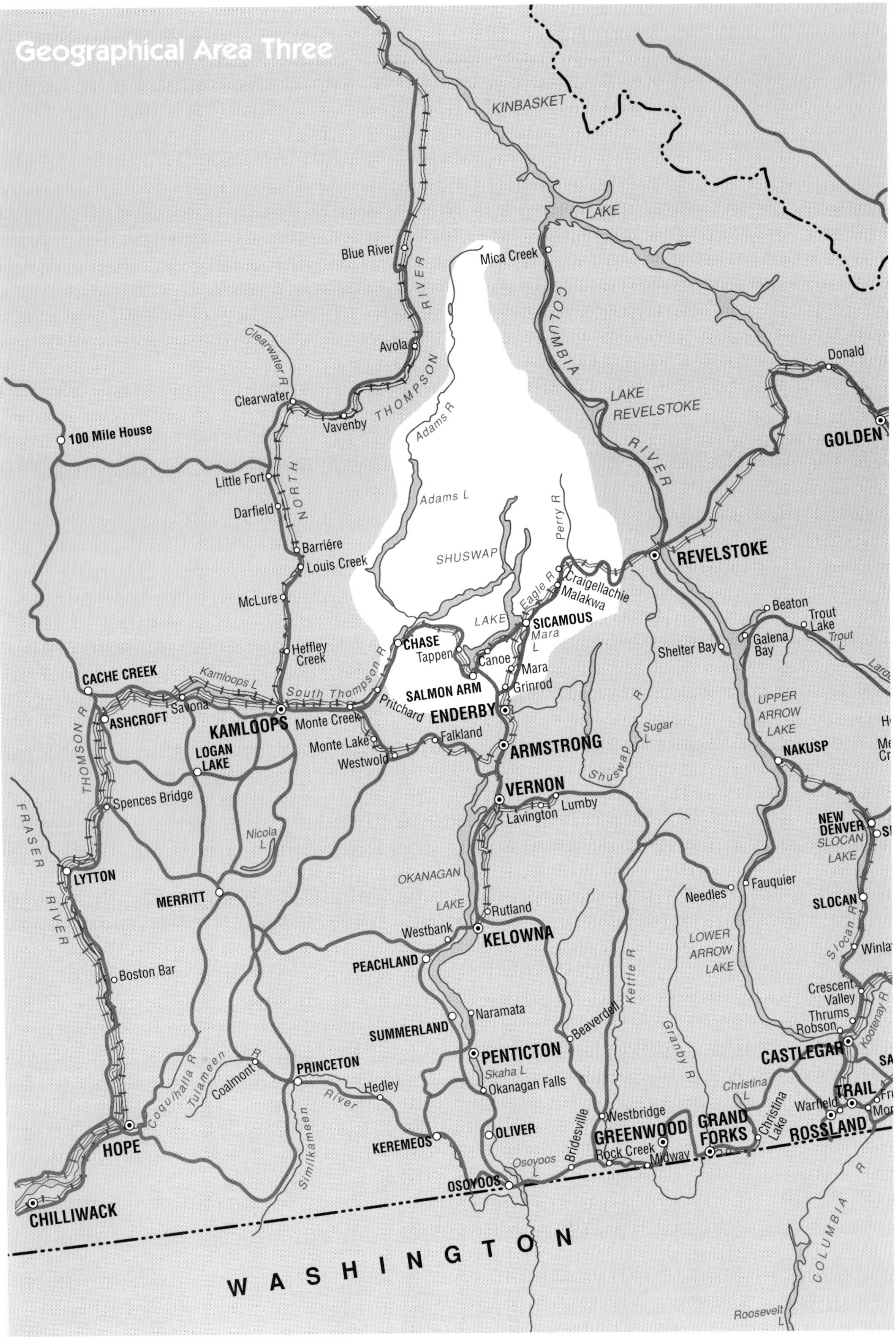

Geographical Area Three
KINBASKET
LAKE
Mica Creek
Blue River
COLUMBIA
RIVER
THOMPSON
Avola
Clearwater R
Clearwater
Vavenby
Adams R
LAKE REVELSTOKE
Donald
GOLDEN
100 Mile House
Little Fort
Darfield
NORTH
Adams L
Perry R
Barriére
Louis Creek
SHUSWAP
REVELSTOKE
Craigellachie
Malakwa
Eagle R
McLure
LAKE
SICAMOUS
Beaton
Trout Lake
Galena Bay
Mara L
Heffley Creek
CHASE
Tappen
Canoe
Shelter Bay
CACHE CREEK
Kamloops L
South Thompson R
Mara
Grinrod
SALMON ARM
Savona
ASHCROFT
Pritchard
ENDERBY
UPPER ARROW LAKE
KAMLOOPS
Monte Creek
Falkland
LOGAN LAKE
Monte Lake
Westwold
ARMSTRONG
Sugar L
THOMSON R
NAKUSP
Shuswap R
Spences Bridge
VERNON
Lumby
Lavington
FRASER
NEW DENVER
SLOCAN LAKE
Nicola L
LYTTON
OKANAGAN
LAKE
MERRITT
Needles
Fauquier
Rutland
SLOCAN
RIVER
Westbank
KELOWNA
LOWER ARROW LAKE
Slocan R
PEACHLAND
Kettle R
Boston Bar
Crescent Valley
Naramata
Thrums
Robson
SUMMERLAND
Beaverdell
Granby R
CASTLEGAR
Kootenay R
PENTICTON
Skaha L
Coquihalla R
Tulameen R
Coalmont
PRINCETON
Okanagan Falls
Christina L
Hedley
River
TRAIL
Warfield
Westbridge
Bridesville
Christina Lake
Similkameen
OLIVER
GREENWOOD
GRAND FORKS
ROSSLAND
HOPE
KEREMEOS
Rock Creek
Midway
Osoyoos L
OSOYOOS
CHILLIWACK
COLUMBIA R
WASHINGTON
Roosevelt L

Geographical Area Three
Shuswap/Adams Lake

R.W. Bruhn Mill at Canoe, B.C.

The first recorded sawmill venture at Canoe was built by R.W. Bruhn in 1925. The mill was operated by A.M. Stewart from 1926 to 1937, when it was sold to Shuswap Lumber Co., owned by Ormie Harris, a founding member of the ILMA. Harris became the second Chairman of the ILMA (1945-46). In 1938 the mill was destroyed by fire but was soon rebuilt and continued operating until 1945.

When economic times improved, fired by the post-World War II building "boom", a ready market opportunity emerged through the formation of co-operatives in the prairie provinces, mainly in Saskatchewan. These co-operatives were interested in reliable sources of lumber. They also wanted to strengthen their member-owned institutions.

One Saskatchewan cooperative, Federated Co-operatives Limited (FCL), seized the opportunity to supply member needs for lumber by purchasing Shuswap Lumber Company in October, 1945.

Shuswap Lumber Co. Ltd. Canoe, B.C. mid 1940's.

Chase Lumber Co. Ltd.
Ormie Harris

Ormie Harris, owner of the CHASE LUMBER CO. LTD. at Chase, rented the mill's facilities to BOB MCKAY in the late 40s. In 1950, GEORGE and JOHN AULIN took over the rental and produced lumber in their company name, NISKONLITH TIMBER PRODUCTS LTD. The Aulin brothers also ran a logging operation.

In 1955, Federated Co-operatives Limited bought the Chase mill. Destroyed by fire in May of 1956, it was immediately rebuilt and back in operation by September of that year.

Bear Creek log flume of the Adams River Lumber Co. Chase, B.C. prior to 1930. (Photo courtesy Kamloops Museum)

Adams River Lumber Co. Chase, B.C. 1919. (Photo courtesy Kamloops Museum)

The Quota System and Consolidation

Companies like Federated Co-operatives Limited, and many others, were profoundly affected by the introduction of the quota system in the late 1950s. Quotas led to many smaller mills amalgamating with larger companies. A company's quota, or volume of allowable cut, was based on the average volume of crown timber cut annually, from 1957 to 1959, by a given operator.

The quota itself was not deemed to have a direct dollar value, but the importance of securing quota obviously had a significant effect on the industry. The introduction of a quota system allowed companies to shift their concentration from finding timber supplies to longer-term planning.

Throughout the interior, the introduction of the quota system, in concert with other changes, caused a period of active trading in smaller mill operations and the eventual consolidation of holdings into larger companies.

Federated Co-operatives Limited

By acquiring several mills during the 1950s and 60s, Federated Co-operatives Limited (FCL) enlarged its quota and expanded its presence in the interior. Its first B.C. purchase, the Shuswap Lumber Company in Canoe, was in 1945.

Art Lloyd came from the coastal forest company, MacMillan Bloedel, to manage the Canoe operation in 1952. He stayed until 1970, when he moved north to Mackenzie to manage sawmill operations for Finlay Forest Industries Ltd. He was President of the ILMA in 1960 and '61.

Fred Stinson was responsible for forestry and logging at the Canoe operation between 1957 and 1966. He was President of the ILMA in 1978 79.

Marv Persson, ILMA President 1976-77, replaced Art Lloyd as General Manager from 1970 to 1978. Charlie McCarthy, from Salmon Arm, replaced Marv Persson upon his retirement and is now FCL's Senior Vice-President – Forest Products. Since 1978 McCarthy has served as an ILMA Director.

Other key players in the Federated Co-operatives Limited enterprise included Ken Seaman who was maintenance foreman and plant engineer until the mid-1960s before going into sawmill engineering private consulting. Al McMath, who was sawmill superintendent at both Chase and Canoe through the 50s and early 60s, left to manage the Rutland and Olinger Lumber Co. Ltd., at Carmi. Bill Kenner, manager of the Canoe Sawmill since 1978, has been an active ILMA Quality Control Committee member. Stu Ross, sales manager for the last 12 of his 40 years with FCL, retired in 1992.

Federated Co-operatives Limited Chase Mill. 1955

Holding Lumber Co.
Art Holding and Jack Hayhurst

Holding Lumber Co. Adams Lake, B.C. Mill startup. 1945

Same operation 13 years later.

From small beginnings, Holding Lumber Co. came to figure prominently in the expansion and modernization of the interior forest industry.

Art Holding had made his start in the lumber business in 1940, as a logger for the S.M. Simpson operations. He and his partner, Jack Hayhurst, bought into a mill on Adams Lake, and formed a wider partnership with Frank Tarry and Sons Ltd. After their mill was destroyed by fire in 1945, Art Holding bought out his partners and rebuilt as sole owner of the company.

Holding was one of the original partners in Kamloops Wood Conversions Ltd., which led to the formation of Kamloops Pulp and Paper Company, and later to Weyerhaeuser's pulp mill in Kamloops.

In 1971, Art Holding sold his own company to Whonnock Industries Ltd, later acquired by International Forest Products Ltd. (Interfor). At the age of 90, Art lives in well-earned retirement near Scotch Creek on Shuswap Lake.

Industrial Timber Ltd.
Art and Mary Nakamoto

INDUSTRIAL TIMBER LTD. was a family operation with a successful but short existence. It was one of many small mills sold to Federated Co-operatives Limited during the consolidation period that followed the introduction of mill quotas.

H.K. Lumber Co. Ltd.
Pete Heron

A mill owned by PETE HERON, located at Pritchard, west of Chase, was also bought by Federated Co-operatives Limited in 1964. F.C.L. later built and operated a plywood plant at Canoe under the name H.K. LUMBER CO. LTD. At the close of the H.K. sawmill, the building was dismantled piece-by-piece, then moved and re-assembled in Smith, Alberta, north of Edmonton.

Kalke and Lawrence; Blanc Brothers; Salmon Arm Planing Mills Ltd.

These were three other small companies that operated for a time in the Salmon Arm area. Kalke and Lawrence and Blanc Brothers were acquired by F.C.L. Salmon Arm Planing Mills Ltd. was acquired by various owners, as already indicated.

Salmon Arm Lumber and Timber Ltd.
Charlie Nakamura

This operation, started in the 1950s, was out of business by 1962. Its quota was split between Federated Co-operatives Limited, Riverside Forest Products Ltd., Cooke Chardon Ltd., and others. As well as a sawmill in Salmon Arm, the firm also had operations at Mabel Lake and Revelstoke.

Shuswap Timbers Ltd.
Jack Carmichael

Jack Carmichael arrived in Canoe in 1945 to manage SHUSWAP LUMBER CO. for Ormie Harris. After Harris sold to FEDERATED CO-OPERATIVES LIMITED, Carmichael stayed on as manager for seven years. With partner COLIN SHANTZ and his son BUD, Carmichael formed SHUSWAP TIMBERS LTD. They acquired Tree Farm License 33 and in the 50s and 60s established a modern sawmill, planer mill and dry kiln before also selling to Federated Co-operatives Limited.

After the sale, Carmichael and Colin Shantz retired while Bud Shantz became general manager for B.J. CARNEY LTD., the pole company at Enderby.

Eagle Pass Timber
Walter Enquist

Eagle Pass is the narrow divide between the drainages of the Fraser and Columbia Rivers, west of Revelstoke. The Trans/Canada Highway and CPR mainline follow the Eagle River valley as the river flows west to join the south Thompson system.

During the 50s and early 60s, WALTER ENQUIST operated a small sawmill in Yard Creek in the Eagle Pass, and a nearby planer on Highway 1. In 1964 the operation was sold to S.M. SIMPSON LTD. and closed down.

McLean Sawmill – Prince George then Malakwa

SINCLAIR MCLEAN, a former Nova Scotian, with sons STAN, SAM and later, NEIL, owned and operated a sawmill in 1938 near Shelley, east of Prince George. They built a second mill nearby and in 1949, moved the first mill to Malakwa. After the Malakwa mill was sold to two Americans in 1956, SANDY PERRY, the company accountant, stayed on as manager.

The next year the new owners sold to S.M. Simpson Ltd. which closed the mill and moved peeler logs by rail from Malakwa to the plywood mill in Kelowna. Sandy Perry remained in charge for S.M. Simpson until the logging operation was shut down in the 60s. Neil McLean established a forestry consulting business in Kamloops after the sale of the family business and, with members of his family, is still active as a consultant.

The Interior Wet Belt Forest "Overmature Decadent Stands"

From the industry's beginning, it was difficult for many mills in Shuswap and West Kootenay regions to operate in overmature timber stands. "Rotten hemlock" was the bane of many operators. A major species in interior wet belt forests, hemlock in some of the original stands was too rotten to be logged economically. The species was also heavy and hard to move. A dilemma was posed because much of the rotten hemlock occupied some of the most productive growing sites in the province.

The sight of these overmature forests can be deceptive. Trees are often huge, giving the impression of great volumes of standing timber. Some old hemlock trees are so rotten they literally collapse when felled, leaving mountains of unusable fiber on the forest floor. The trees might actually still be alive, but when the annual volume of decay exceeds the growth volume, the stand is considered "decadent".

Throughout B.C.'s debate over old growth forests, the term "decadent" became sharply controversial. Differences in definition created fundamental disagreement. Defined by the forestry community, decadent is a technical term applied to stands in which decay was accumulating faster than new wood. Advocates of old growth preservation believed the term "decadent" was used to justify logging, usually by clearcutting, of old growth stands. They considered it a dismissal of the intrinsic value of old growth.

At the introduction of sustained yield policies several decades ago, government foresters urged companies to log decadent stands as soon as possible, in order to get healthy young forests established in their place. The decision to proceed with logging was a technical call, based on an understanding of the future growth potential of these sites.

Loggers knew that old hemlock stands were dangerous places in which to work. Contractors were not exactly lining up for a chance to cut them. Records on the number of tree fallers killed or injured in rotting hemlock stands are unavailable. But today's fallers, with experience in these dangerous working conditions, believe the accident rate must have been high.

It is ironic, in light of the movement to preserve old growth, that companies logged in some of these overmature stands with great reluctance. In some cases, even if companies were given the right to cut the trees for nothing, they could not break even, much less show a meagre return for their work. A list of companies which found themselves in financial difficulty after working in old growth hemlock, might be quite long.

The Forest Service's ongoing program to remove rotten hemlock is now mandated by new regulations on reducing waste left in the bush. Rotten hemlock logs are removed whenever possible. Even if useless for lumber, when chip markets are strong they can be hauled to a chipping plant in Cache Creek.

Drew Sawmills Ltd.
Ed Drew and Ray Fujikawa

In 1951 ED DREW built a portable mill to process timber cut on the family ranch, west of Stump Lake. In 1969, Drew and his partner, RAY FUJIKAWA, bought the PERRY RIVER LUMBER CO. mill at Malakwa, between Sicamous and Revelstoke, from MORRIS LERMAN and ESMON LANDO. They supplied their mill with wood from the Perry River Timber Berths and Crown timber sales, with much of the volume originating from decadent cedar/hemlock.

JIM MCKINNEY had purchased a nearby mill from the Merritt Diamond group in 1961. After operating for 15 years, it went into receivership and was picked up in 1977 by DREW and FUJIKAWA. Poor markets and high interest rates in the early 80s' recession forced Drew Sawmills out of business.

Ed Drew, who was President of the ILMA (1973-75), retired. Ray Fujikawa, along with partner PAT COADY formed OKANAGAN LUMBER SERVICES LTD., a company that operated in Enderby from 1985 to 1993. They manufactured bevel siding and tongue and groove panels using 100 percent interior cedar.

It's A Risky Business. It Always Was.

Smaller interior sawmills had to be wary of a host of operational uncertainties and pitfalls, any one of which had the potential to cause a shutdown. Circumstances and situations were often beyond the control of owners and managers. Sudden or severe drops in export markets or prices, or lack of an adequate or sustained log supply could choke a small lumber company out of business. And there was the ever-present risk of fire.

Some mills lived a charmed life, running for years with the ability to avoid major obstacles. Others ran into every obstacle. The history of some mills runs the gamut of unsettling experiences, from fire to bankruptcy, to interrupted wood supply, to changing ownerships and partnerships.

Some mills have such checkered and complicated histories, they can best be recorded by highlighting the major changes. Malakwa Sawmills Ltd. is presented as an ideal candidate for a highlighted version, as outlined in the following section.

Malakwa Sawmills Ltd.

Dan Reiswig and his sons Harvey and Ed opened a mill in 1956 to custom saw wood from the Perry River timber berths owned by Morris Lerman and Esmon Lando. Three years later the mill was destroyed by fire and immediately rebuilt.

1961 Reiswigs sold to Merritt-Diamond Mills Ltd., who shortly after sold to Jim McKinney.

1977 McKinney's Malakwa Sawmills Ltd. went into receivership and was purchased by Drew Sawmills Ltd.

1983 Drew Sawmills went into receivership.

1985 The mill was bought from the receivers by Mel Beaumont who picked up the old Perry River Lumber Co. at the same time. He called his company Beau J. Mills Ltd.

1988 The Perry River mill was bought by Westar Timber and became known as the Eagle River operation.

1991 The old Reiswig mill was converted to a special products operation and continues operating at Malakwa under the name Beau J. Mills Ltd.

1993 Westar sold the Eagle River operation to Evans Forest Products Ltd.

After selling their original mill to Merritt-Diamond in 1961, the Reiswig family continued to operate various sawmills and stud mills over the years.

Loading logs in the early Shuswap years. (Courtesy Salmon Arm Museum)

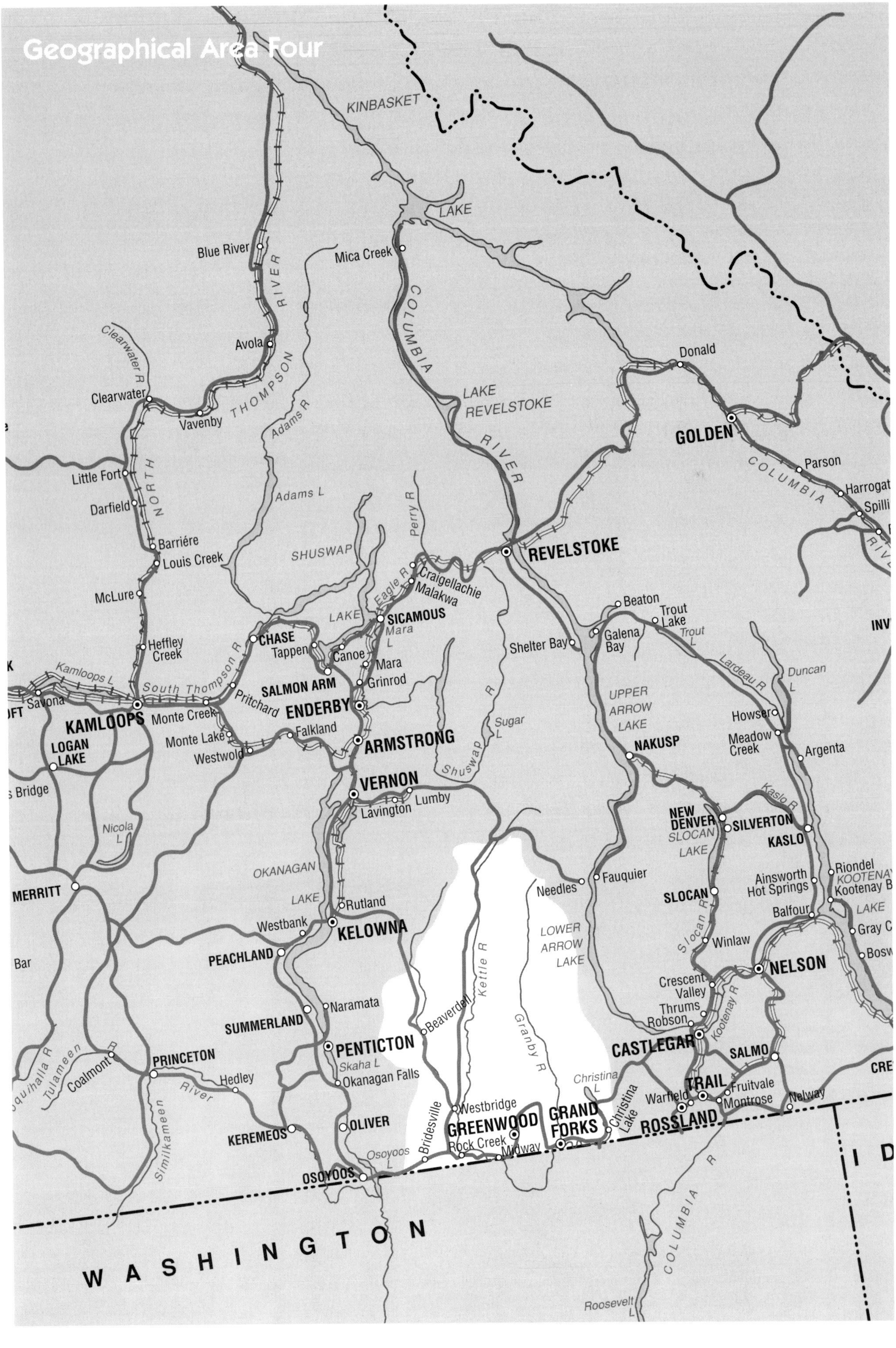
Geographical Area Four
KINBASKET
LAKE
Mica Creek
Blue River
RIVER
THOMPSON
NORTH
Avola
Clearwater R
Clearwater
Vavenby
COLUMBIA
LAKE
REVELSTOKE
RIVER
Donald
GOLDEN
COLUMBIA
Parson
Harrogat
Spilli
Adams R
Adams L
Little Fort
Darfield
Barriére
Louis Creek
McLure
Heffley
Creek
SHUSWAP
Perry R
LAKE
Eagle R
Craigellachie
Malakwa
REVELSTOKE
SICAMOUS
Mara
L
Mara
Grinrod
CHASE
Tappen
Canoe
SALMON ARM
Kamloops L
South Thompson R
Savona
KAMLOOPS
Monte Creek
Pritchard
ENDERBY
Falkland
Monte Lake
Westwold
ARMSTRONG
LOGAN
LAKE
Bridge
Nicola
L
VERNON
Lavington
Lumby
Shuswap R
Sugar
L
Shelter Bay
Beaton
Trout
Lake
Trout
L
Galena
Bay
Lardeau R
Duncan
L
UPPER
ARROW
LAKE
NAKUSP
Howser
Meadow
Creek
Argenta
Kaslo R
NEW
DENVER
SILVERTON
SLOCAN
LAKE
KASLO
Riondel
KOOTENAY
Kootenay B
INV
MERRITT
OKANAGAN
LAKE
Rutland
Westbank
KELOWNA
PEACHLAND
Needles
Fauquier
LOWER
ARROW
LAKE
SLOCAN
Ainsworth
Hot Springs
Balfour
LAKE
Gray C
Bosw
Slocan R
Winlaw
NELSON
Bar
Kettle R
Granby R
Crescent
Valley
Thrums
Robson
Kootenay R
Naramata
Beaverdell
SUMMERLAND
CASTLEGAR
PENTICTON
SALMO
PRINCETON
Skaha L
Okanagan Falls
Christina
L
Tulameen
Coalmont
Hedley
River
Similkameen
Bridesville
Westbridge
GREENWOOD
GRAND
FORKS
Christina
Lake
Warfield
TRAIL
Fruitvale
Montrose
Nelway
ROSSLAND
OLIVER
KEREMEOS
Rock Creek
Midway
Osoyoos
L
OSOYOOS
COLUMBIA R
WASHINGTON
Roosevelt
L
I D
CRE

52

Geographical Area Four
Kettle Valley/Boundary District

Forest licensing and tenure arrangements have been serious business in a province with 95 percent of its land base owned by the public. Successive B.C. governments have tried various approaches to regulating resource users. Finding the right fit has called for the wisdom of Solomon.

Governments, and their commissions and task forces, have reviewed and analyzed, investigated and surveyed to devise forest tenure and licensing arrangements that provide the greatest social, environmental and economic benefits. (Other government jurisdictions, notably in the United States, have not faced nearly as intense a problem since most of their forest land is privately owned.) At times when various B.C. regulatory agencies and government Ministries have had conflicting agendas, as they often have, complications have grown worse.

Admittedly, B.C.'s policy-makers have faced a formidable challenge. As well as protecting the forest resource itself, governments have had to derive revenue to provide education, health and transportation services. Stability and order have been needed within the industry, and communities have had to be sustained.

The highwire balancing act between these objectives in dynamic social and economic conditions has been an unenviable job. Despite the best-laid plans of government, trade-offs between benefits and costs have often been unavoidable. One change in forest tenure or licensing, heralded as progressive by some groups and individuals, was bound to be attacked by others.

Unfortunately, when government changed one policy after recognizing a problem, the change often created more unanticipated problems and tensions. A shift in policy intended to provide long-term benefits could precipitate an immediate crisis.

Government regulators were not intentionally oblivious to the needs of timber-dependent communities. But once in place, policy tended to be inflexible and slow to respond to changing situations. The diverse geography and climate of regions virtually ensured that policies suited to conditions in one region would be a poor fit in another.

Inertia characterized government process. By the time regulators identified a problem and changed a policy, circumstances had frequently changed again, and the new policy was out-of-date. The inability to stay current was especially problematic in the ultra-dynamic forest environment of the 1960s. To some extent, the inertia has been passed on, and sometimes intensified, to the present.

Forest companies, meanwhile, had to stay flexible to adapt to the fast pace of change in licensing or tenure. As soon as they grew accustomed to one system, it was often replaced with another. Complex changes in arrangements were especially daunting for owners and managers of many small firms. Although they were highly capable at the mechanical level, many lacked the skills, the time or the inclination to cope with bureaucratic regulation and increasing amounts of red tape.

Particular forest licensing policies stand out in history as highly controversial. The well-intentioned timber auctions of the 50s and 60s were a case in point. They proved to be as controversial and disruptive as any government initiative.

The Bidding Wars

The pitfalls of open bidding on timber were apparent soon after the first series of auctions. Prices for many Crown timber sales were bid to unrealistically high levels. Many mills found, to their alarm, that a successful bid gave them harvesting rights to timber that was no longer economical to harvest. Out of desperation, forest companies often stayed in the bidding even when prices pushed profit margins into the red, just to keep their mills open and their crews working.

The open bidding system, ostensibly designed to allow the free market to determine the price of logs, was based on the principle that the public should be given equal opportunity of access to cutting rights. In practice, this system had major flaws. Workers at a local sawmill could lose their jobs overnight if their employer lost an important wood supply to an outside bidder. A heavy social price had to be paid when a mill went down. The loss inevitably resulted in severe disruption of families and sometimes entire communities.

The open bidding system also provided exclusively for short-term timber demand. It offered no incentive for companies to plan for or invest in the long-term health of the forest or their businesses.

Under the bidding system, individual operators could apply to the forest service for cutting rights in a given area. They would first have assessed the timber stand to determine if it was suitable for logging. The government then served notice to the public that the timber was for sale. Anyone could pay the required deposit and enter a bid. Bidders did not have to be from the area, or even own a sawmill to be successful at getting cutting rights to the timber. Original applicants, usually local sawmill owners, could end up without any wood supply from the forests in their own region.

The negative consequences of the open bidding system soon reached beyond the cool level of business. Companies and individuals, set up in heated confrontation for their livelihood, sometimes took the loss of important timber sales personally. Spite and grudge bidding built tremendous resentment and created a hostile feuding atmosphere.

Meanwhile, sales of Crown timber at exorbitant prices did not prove to be the economic windfall to the government that might have been anticipated. Although stumpage rates were bid through the roof in some cases, the overall system did not guarantee high revenue for the province. Many forest companies went broke or were forced to sell out when their operations were no longer viable. The resulting social dislocation and disruption in local economic activity often translated into more government expenditure than revenue.

The unstable business climate fostered by the open bidding system resulted in some strange alliances between competing companies. This was the case with K-V Timber Ltd.

K-V Timber Ltd.

In the late 1950s the Forest Service administered PUBLIC WORKING CIRCLES. PWC's were relatively large geographic areas in which the Forest Service issued timber sales through public auction.

In a move to defuse the "spite and grudge" bidding atmosphere, a group of forestry firms in the Kettle Valley formed an unusual alliance, their own version of a working circle.

Tony Hilliard, a forest engineer with Boundary Sawmills Ltd., spearheaded the formation of a loose association of forest companies. The association included Boundary Sawmills Ltd., Cooke Lumber Co., Fritz Sawmills Ltd., Carlson & Finch Lumber Co. Ltd., Mountainside Sawmills; and Willa Lumber Co. Ltd. Together the companies formed K-V Timber Ltd. and hired Registered Professional Forester, Gerhard Eichel, to coordinate the company's affairs, effective January 1, 1960.

Forest Service policy at the time stipulated that, when there were no competing bidders for a cutting opportunity, the original applicant could acquire the timber for an "upset" price. The "upset", based on the Forest Service's estimate of the value of the standing timber, was the price at which bidding had begun.

K-V Timber's singular purpose was to monitor timber auctions. It would not enter a bid unless a company from outside the operating area submitted a bid against a local operator. Although Eichel was strategically poised to enter the bidding at any time during K-V Timbers' ten-year lifespan, he never once had to exercise this option. Perhaps the mere existence of the K.V. Timber alliance was enough to keep outside bidders at bay.

Predictably, when K-V Timber was first incorporated there were angry protests that its presence would restrict the free functioning of the market place. Under the law of the day, it was a legal activity that marked an interesting, if volatile period in the history of logging and sawmilling in the Boundary country.

K-V Timber Ltd. was dissolved when the open bidding process was replaced in the late 1960s by the issuance of timber sale harvesting licences. Refinements to this licensing system have been ongoing.

K-V Timber forester Gerhard Eichel was hired by Boundary Forest Products Ltd., later Pope and Talbot Ltd., where he worked until retirement in 1990.

Boundary Sawmills Ltd. Sherbinin and McMynn Era

This Depression-era forest operation hand-logged with axes and cross-cut saws and skidded logs with horses.

In 1935, John Sherbinin and his son John Jr., with other relatives and friends, started a sawmill north of Midway at Ingram Creek. The mill cut fir, larch and pine ties. Other timbers were sold, primarily to the CPR, at a going rate of $10 per thousand board feet. The steam-powered and wood-fired mill could cut 35,000 board feet of lumber daily. (A three-bedroom frame home requires approximately 15,000 board feet.)

Pete, Bill, John Jr. and J.W. Sherbinin (father) 1943

Boundary Sawmills at Ingram Creek. 1936

The Sherbinin family also ran a cook-house during the Depression years, buying its supplies from the grocery store in Midway. The store's owner, Gordon McMynn, took his first direct step into the lumber business by "grub-staking" – providing grocery goods to the cook-house in return for a share in the lumber operation. The grocer-turned-lumberman eventually became a full-fledged partner in the enterprise. McMynn served as ILMA Chairman in 1947-48.

In 1936 the Sherbinins and McMynn purchased a small sawmill in Midway that produced lumber for shipment by rail to the prairie provinces. By 1940, combined production from the Midway and Ingram Creek operations was at 9 million board feet annually. These operations were incorporated in 1943 as BOUNDARY SAWMILLS LTD.

During the war years, Japanese families were relocated from the Coast into the Midway and Greenwood area and many found work in the mills and logging operations. SEIJI ONIZUKA worked for the company and its successors for more than 40 years.

Boundary Sawmills . 1957.

When J.W. Sherbinin died in 1946, his son JOHN JR. became President of the firm. After the Ingram Creek mill burned in 1950, the Midway mill, located on the present site of the Pope and Talbot mill, was enlarged and modernized.

Boundary Sawmills continued to grow to an annual production of 30 million board feet. It was granted a Tree Farm License in the early 1950s, thanks to the work of GEORGE MINNS and TONY HILLIARD. When John Sherbinin Jr. died in 1953, Gordon McMynn succeeded him as President, a post he held until his death in 1966. Third generation GEORGE SHERBININ then became President, and his brother Gus, took over as Sales Manager.

Following the pattern set by mills elsewhere in the province, Boundary Sawmills grew by buying out and consolidating a number of local operations. It expanded its operating territory with the acquisition of GRAND FORKS SAWMILLS, FRITZ SAWMILL LTD., OLINGER SAWMILLS LTD. and the timber quota of SANDNER BROTHERS LUMBER CO. LTD. at Christina Lake. The consolidated enterprise was renamed BOUNDARY FOREST PRODUCTS LTD. After the protracted strike of 1967, and faced with the costs of expanding and modernizing to utilize "third band" timber, Boundary Forest Products was sold in 1969.

Enter Pope & Talbot Inc.

Pope and Talbot Inc. of Portland, Oregon became the new owners of Boundary Forest Products in 1969 and later renamed the Canadian operations POPE AND TALBOT LTD.

Pope & Talbot Ltd. has a unique niche in the interior industry. A major part of its log production and most road construction and maintenance are carried out by company employees. The industry norm is to contract out many of these operations.

George Sherbinin resigned as President of Pope & Talbot Ltd. in 1974. He continues his family's long history of involvement with the forest industry as President of S & O SAWMILLS, a mill and planer operation north of Westbridge and as owner of GEORGE SHERBININ LUMBER LTD., a wholesale firm in Midway. Replacing George as President was DICK REISER, an American citizen. Dick retired in 1987 at which time ABE FRIESEN was appointed President. Abe Friesen is currently a Director of the ILMA.

Other well known figures associated with the company were ARCHIE JOHNSON, a colourful "bull-of-the-woods" character who worked as logging boss at Ingram Creek and later out of Midway until his retirement. TONY HILLIARD, a UBC forest engineer, was hired by Boundary Sawmills Ltd. in 1951 and remained with the company when it was taken over by Pope and Talbot Ltd. Hilliard left B.C. in 1972 to manage a new Pope and Talbot acquisition at Sioux Lookout in northern Ontario. He later returned to join Weyerhaeuser Canada Ltd. in Kamloops, where he stayed until retiring in 1983. Another well-known figure in this company's operations, BILL UPHILL, was a former forest ranger who joined Boundary Forest Products in 1969. Uphill retired as vice-president of Pope and Talbot Ltd. in 1990.

Pope and Talbot Ltd. expanded again in 1992 with the purchase of WESTAR TIMBER'S sawmill at Castlegar, and its timber supply in the Arrow Lakes.

Norris Lumber & Box Factory Co. Ltd.

At Grand Forks, the Boundary-country town at the junction of the Kettle and Granby Rivers, Norris Lumber and Box Factory operated in the 20s and 30s. Founder Ben Norris was active in the company until it was sold to five Okanagan cooperatives with an interest in a dependable supply of fruit boxes. The new owners renamed the company Grand Forks Sawmill Ltd. Boundary Sawmills bought out the cooperatives in 1957, and continued to operate the mill under the same name. In 1967 the sawmill purchased major holdings in Sandner Brothers Lumber Co.

Cooke Lumber Co.

Ed Cooke

In 1950 Ed Cooke traded an old tractor for a portable sawmill a short distance from Greenwood. When fire destroyed the mill, Cooke built another mill in Greenwood. Cooke Lumber Co. expanded in 1955 with the purchase of Rock Creek Forest Products from the McDiarmid interests. In 1962 Cooke bought Mountainside Sawmills Ltd. from Crae Dawson. And in 1963, he added to his thriving enterprise Chardon Lumber Co. Ltd. in Revelstoke; previously owned by Charlie Nakamura.

Cooke Lumber Co. Ltd. Greenwood, B.C. early 1950s

Cooke's Revelstoke mill operated as Cooke-Chardon Lumber Co. Ltd. until 1965. The site was then sold to B.C. Hydro to make way for construction of the High Arrow (now Keenleyside) dam. Cooke sold out to Northwood in 1967 and worked for that company until retiring in 1971. He died in 1988.

Kettle River Sawmills

Gerard Lautard

Kettle River Sawmills has a history that is typical of the era of small mills. In the spring of 1947, Gerard Lautard, was in his early 20s. The son of a pioneer West Kettle family, Lautard bought a mill and power unit for less than $2,000. He ran his mill and logged as a one-man crew, until 1950 when he hired one helper. In 1954 he moved the mill from Wildhorse Creek to a site on Highway 33 at Rhone, its present location. Kettle River's main products were rough timbers and crossing planks for the CPR.

The company bought a planer in Osoyoos in 1974, and operated it there for three years before transferring the planing to the millsite at Rhone. With the planer acquisition the mill became a member of the ILMA. It received the ILMA's zero-frequency safety awards in 1975, 76 and 79.

Lautard sold out in 1979 to Hillmoe Forest Products of Bridesville. In 1983 the mill was acquired by S & O Sawmills. The following year S & O Sawmills hired Lautard out of semi retirement as a sawyer. After 5 years with S & O, Lautard permanently retired to Penticton.

Sandner Brothers

Lincoln (Linc) and Randolph (Rand)

Linc and Rand Sandner started a small shingle mill and sawmill on the south end of Christina Lake in 1936. During the war years, their business expanded with healthy infusions of capital and with the help of a reliable labor force. They hired many of the Japanese Canadians who had been relocated to the Interior. Many of these hardworking employees stayed on at Sandner's operations until they retired.

Sandner Brothers started a box plant in Greenwood in 1944. This part of the operation closed in 1970 when the fruit industry converted to paper boxes. In the late 50s the steam-powered mill at Christina Lake was converted to electric power. It replaced its circular head saw with a band saw rig, and then with a gang saw in the early 1960s.

In 1967 the Sandners struck a deal to process Northwood's timber quota out of the Salmo Sustained Yield Unit. Northwood would then sell the lumber. The arrangement worked well for both parties and equipped Sandners with the resources to install a barker and chipper in 1971.

After Northwood's sale to Weyerhaeuser in 1979, Sandner Brothers Lumber continued to cut the Salmo quota. In 1983, that arrangement ended when Weyerhaeuser's quota allotment from the Salmo management unit was transferred to the Okanagan. A year later, after 48 years of nearly continuous operation, Sandner Brothers Lumber closed down.

Other Boundary Operations

The Boundary area experienced a proliferation of sawmill operations of all sizes between the post-war years and the late 1960s. These operations varied in size and length of operating life and most eventually disappeared as separate entities with the acquisition or consolidation activities of others.

Among the mills and millowners that passed into Boundary history were OLINGER LUMBER of Carmi, BERT ELLET LUMBER CO. of Beaverdell, ALF ACKERMAN at Taurus, LEWIS CHARD, CLIFF FILLMORE and KETTLE RIVER SAWMILLS at Rhone, MILLER LUMBER CO. at Westbridge, PETE BOSOVICH, HELGE FROM and WILLARD TRAINER on the Main Kettle River, CHARLEY WILSON at Hume Creek, the SHIPPIT BROTHERS, ANSCHETZ & BOHNET, OLIVER VALLEE, CARL BOURGH in the Rock Creek area, and ART ALDEN in McKinney Creek, BILL HILL at Sidley, BOUNDARY FALLS LUMBER (south of Greenwood), BILL AND ALEX PADMOROFF in Norwegian Creek, LAWRENCE FOLVIK in McCarron Creek, JIM AND PAUL FORSHAW near Greenwood, BRYANT & SIMPSON at Brown Creek, K.P. DONDALE at Pass Creek and ERIC ATWOOD & PETE POPOFF at Morrisey Creek. There were many others.

Close Utilization and "Third Band" Wood Transform the Interior Industry

The harvesting standard of the early interior forest industry was termed "rough utilization". Only large diameter sawlogs could be economically harvested at the time.

As technology progressed, rough utilization gave way to an "intermediate utilization" standard that made greater use of the timber resource.

During the 60s, the B.C. government was intent on further tightening of utilization standards. There was strong incentive to do so. Increased revenue and employment would certainly result if more wood volume could be utilized. A search was on to find ways to utilize the substantial portion of the forest that had traditionally been considered uneconomical to recover.

"Close utilization", the forestry buzz-phrase of the day, introduced the change that was to transform the interior industry. It meant that more wood had to be recovered from the same area than had been the practice in the past. This was to be accomplished by lowering the height of stumps (the most valuable part of the tree being at the bottom), taking the tops down to a 4-inch diameter and harvesting smaller or overmature trees, rather than leaving them behind.

In a speech in the Okanagan in October, 1968, B.C.'s Minister of Lands, Forests and Water Resources, Ray Williston decried the standards of the time that were "resulting in excessive waste of sound-wood volumes which would be avoided by the adoption of close utilization practices."

Williston referred to the results of detailed studies which showed great increases in usable wood volume with moves from rough, to intermediate, to close utilization in the interior forest. The Okanagan area's pine resource was of particular interest to Williston. He reported that "the most dramatic statistics are to be found in those areas in which there is substantial lodgepole pine cover."

Close utilization remained a concept until the time was right. Its actual introduction depended on the demand for more wood and the development of technology that could efficiently produce more volume from the same tree. Market demand spurred the development of this new technology in the late 1960s. The major change to close utilization in the interior forest industry was at last underway.

The logical extension of the close utilization strategy and the promising new milling technology expanded operations into vast stands of lodgepole pine, a species that had not previously been economically viable for harvesting.

"The major change that has happened during my tenure with the industry is what I call the jackpine revolution. During the

1960s there were huge areas of lodgepole pine in the interior forests that probably established as a result of the large fires in the late 1800s. This was the time the industry was developing technology to handle this timber both in logging and sawmilling. And the lodgepole pine, which had been a weed species, became the bread and butter of the industry. Now, in B.C., we log more lodgepole pine than anything else."

– Bill Uphill, (Retired Vice President, Pope & Talbot Ltd.)

Across the interior, forest companies began gearing up to handle this "new" wood. Literally hundreds of millions of dollars in capital were invested into the research, development and installation of new technology. This vast lodgepole resource was made available to companies willing to invest in the new technology. To accommodate the lodgepole, the interior industry underwent a radical transformation.

The additional volume of timber obtained by operating in these previously bypassed stands became known as "third band" wood. (The term originated from a three-banded illustration used by Ray Williston to show the potential for an increase in wood volume.)

Third band wood represented a huge increase in the volume of timber, almost 40 percent of the cut in some areas. Before the introduction of close utilization standards and third band, lodgepole pine had accounted for only about 1 percent of the provincial cut. In the interior more lodgepole pine is now logged than any other species.

John Murray, who worked for Kootenay Forest Products and later, Crestbrook, describes this period as one of the most dramatic in his career, in terms of its impact on forest operations. It brought about sweeping changes in both logging and processing.

Retired Chief Forester Bill Young, a founding member of the B.C. Forest History Association, recalls that there was virtually no demand for lodgepole pine at one time. Some timber sales contained a "log it or leave it" clause that gave the operator the option to cut the pine, instead of selecting just the spruce or fir, or to leave it standing. "Log it or leave it" clauses were phased out and operators were required to log and process the pine with the other tree species.

Some operators stoutly resisted the changes. Others adapted swiftly to the challenge, building mills for the express purpose of cutting the large volume (albeit in small trees) made available when the pine became a "new" resource.

"One of the most important changes in the industry was close utilization. It greatly reduced waste wood, the need for big burners and so on. We went from rough utilization 40 years ago to what we call close utilization today, which is a 6-inch butt and a 4-inch top. Now there are a number of companies that are utilizing the material made available by going to even smaller top diameters."

– John Murray, (Retired Vice President, Woodlands Div., C.F.I.)

Selling Lumber – Selling Mills

With the rapid advancement of technology, doors closed on many old and out-dated mills. Mills equipped with chipping capabilities took their place. These mills met the objectives of close utilization by taking the usable volume of lumber from small diameter trees and chipping the leftovers for pulp.

This cutting edge technology was extremely expensive and heralded another consolidation period, as larger better-financed operations were in the best position to keep up with the changing business environment.

"The consolidation of companies has not necessarily reduced the amount of employment. It has allowed for integration into large companies that are more stable through the market cycle highs and lows. Federated Co-operatives Limited bought a little mill in 1945 at Canoe, and since that time has acquired small companies in Chase, Pritchard, Sicamous and Salmon Arm. Being able to consolidate our manufacturing base and timber supply has permitted very stable operations and employment for 50 years. I don't think small companies could have withstood some of the adversities experienced through the period."

– Charlie McCarthy , (Senior Vice President, Federated Co-operatives Limited)

"At that time Mr. (Ray) Williston was the Minister of Forests. I know that he was pushing for mills to get into third band wood and it meant we had to spend at least

five million bucks at our Midway mill in order to be able to handle wood down to a four-inch top. We had just gone through a seven-month strike and were already in hock for about half a million dollars. In those days that was a lot of money. All over the interior other mills were going through the same thing. They all went for the bigger mills because that was the only way to survive – you just couldn't stand still. So we looked around and eventually got together with Pope and Talbot."

– GEORGE SHERBININ, (PRESIDENT, S&O SAWMILLS LTD.)

"*We started doing small log harvesting in 1982. We took some of the smallest wood in the province, down to the smallest top. That has been a challenge. Most people said it couldn't be done; shouldn't be done. We haven't been too successful at making money at it but we have employed a lot of people and made good products out of it."*

– ERIK NORGAARD,
(PRESIDENT, ARDEW WOOD PRODUCTS LTD.)

"Chip-and-Saw" Mills and The Pulp Connection

Investments in new milling systems were just part of the equation. The government also required companies interested in third band wood to install debarking and chipping systems that would increase the resource for the pulp industry out of slabs and trim ends previously burned.

The introduction of "close U" and third band wood thus created the resource for the interior's chip and pulp industry. While it is true that even a mill specifically designed to handle small diameter wood might get only one or two narrow pieces of lumber out of a small log, the rest of the tree provided an important new volume for chippers.

New "chip-and-saw" mills, though very costly, began to appear throughout the interior. They operated at ultra-high speed compared to conventional head-rig sawmills. Many mills kept their headrig sawmill equipment for large wood and ran another "side" to process the small wood. Usable residue from both milling systems was chipped for pulp mills, reducing burner use, in addition to providing a secure chip supply.

The changeover did not occur without problems. New and introduced to the market in haste, the first chip-and-saw mills were a nightmare of breakdowns and set-backs. Modifications and improvements could ultimately cost more than the mill itself. Without ingenious innovation, not by engineers and scientists, but by the millworkers themselves, the process to smooth-running technology would have taken much longer.

The move to close utilization standards and third band wood literally transformed every level of the interior industry. Technology changed in the woodlands as well. Gradually, feller-bunchers (machines which cut and pile smaller tree stems) replaced many of the handfallers. These machines will never replace handfallers on steep ground or with big trees. But in the right terrain and forest situation, they are now a common harvesting instrument.

Beetle Salvage Logging:
A Natural Event Causes Great Fluctuations In Cut

The volume of wood products manufactured in B.C. rose sharply with close utilization and third band. Nature gave the province another reason to log more trees.

Insect populations are a common and normal part of the lifecycle of western coniferous forests. However, hundreds of thousands of hectares of interior forestland had reached maturity or overmaturity – the forest conditions most vulnerable to insect attack – at the same time. Insects built up to epidemic proportions and huge areas of infested trees were unable to survive.

Two common interior species, spruce and lodgepole pine, were most affected. The sap stain introduced by the mountain pine beetle causes the wood to discolour quickly. Many overseas buyers will not accept the stained wood. Infested pine also cracks or checks unless milled quickly after dying.

It was clear to the government that failure to deal with the infestations would cause an unacceptable waste of the resource. As there were no effective technical or chemical controls for bark beetle outbreaks, the government decided the worst infestations should be removed. Companies throughout the interior were suddenly required to log vast forest stands to prevent the spread of the beetle.

> *"Because these huge lodgepole pine stands were all becoming mature at about the same time, they came under attack by the mountain pine beetle and consequently the industry went up to full speed to utilize the beetle-killed timber. Otherwise there would have been a lot of wasted wood in the area."*
>
> – BILL UPHILL, (RETIRED VICE PRESIDENT, POPE & TALBOT LTD.)

The volumes of timber logged shot up as the Forest Service targeted infested or threatened stands for salvage logging. The rationale was to remove as much of the infected stands as quickly as possible to control the beetles' spread into adjoining stands.

The program to control the spread of the beetle created planning nightmares for the Forest Service and industry. It also caused disruption for agencies responsible for other forest values, such as fish, wildlife and range. Some wildlife species were adversely impacted. Trying to stay ahead of the fast-moving insects caused huge fluctuations in the rate of harvest. Planned operating areas had to be abandoned and new roads built quickly into infested or susceptible stands.

Although there was scientific and economic justification for salvage logging of beetle-kill, the strategy created controversy. Currently close to half the harvesting carried out in the southern interior relates to the control of forest insects and diseases.

As the volume of beetle-killed wood is reduced and the outbreaks run their course, cutting rates will be reduced – at least until the next beetle cycle occurs. Time will tell, but because beetles tend to prefer original and overmature forest stands, the interior's second growth trees may be less vulnerable to infestation.

Logging in Ingram Creek, 1940s.

Dispute over government regulations. Boundary Sawmills, Grand Forks, 1962.

1922 Model T Ford Truck, Black Mountain, Rutland, 1935. (Courtesy Kelowna Museum)

Lighter Side

"In the late 1940s, one of the packing houses phoned us and said they were having a terrible time getting their onions dry. It had been a very wet fall. We agreed to dry several hundred tons of onions in our dry kiln. The first truck driver arrived at the kiln and said to the operator, "Where would you like me to put the onions?" And the operator said, "Onions, what onions? This is a sawmill." They finally got things sorted out and the onions were put in the kiln. Unfortunately, as they dried out, the skins started to pop off. Onions skins were flying out through the vents and the air was permeated with the smell of onions."

– Horace Simpson, (Retired President, S.M. Simpson Ltd.)

"After an ILMA meeting some of us went skinny-dipping. Ed Cooke gave a big war whoop and as he went off the diving board he lost his false teeth in the plunge. We were going for steak dinner afterward so Vern Vance, Ian McDiarmid and I went diving with flashlights to try to find the teeth but to no avail. When we were having dinner the waitress told us she had a son who was a scuba diver. He went down and, lo and behold, he found Ed Cooke's teeth. As a reward, Ed took him on a tour of the Okanagan in his private plane."

– Warren Larson, (Retired General Manager, Bell Pole Company Ltd.)

"When things got slow, we'd go on junkets, driving off into the bush looking for mills. It was easy. You drove on the backroads until you saw a pile of smoke. That meant there was a burner nearby and workers to organize. One afternoon I drove out to a mill, put on my IWA hardhat and walked

around the green chain for a while. After I left, the boss called a meeting and told the workers they were all going to get a 10-cent an hour raise on their next pay cheque. One guy called to thank me for the raise. But he asked me not to come back."

– BILL "WILLIE" SCHUMAKER,
(RETIRED PRESIDENT, IWA LOCAL 1-423)

"I have a buddy – a priest in a local church here. The priest used to come to my office and we would have a drink and then we might end up at his rectory and have a few more drinks. They were going to build a new church and he said, "Will you help us build a church?" I said, "Okay, Father how much do you want?" He said, "You believe in God so you donate the lumber and I will pray for you." It cost me 2,900 dollars. But I am of the Sikh faith. That was not my church."

– TARA GHOG, (OWNER, ASPEN PLANERS LTD.)

Michigan Big Wheels – Used at Canyon, B.C. in 1910.

'Hand Falling' B.C. Spruce Mills Ltd., Lumberton, B.C.

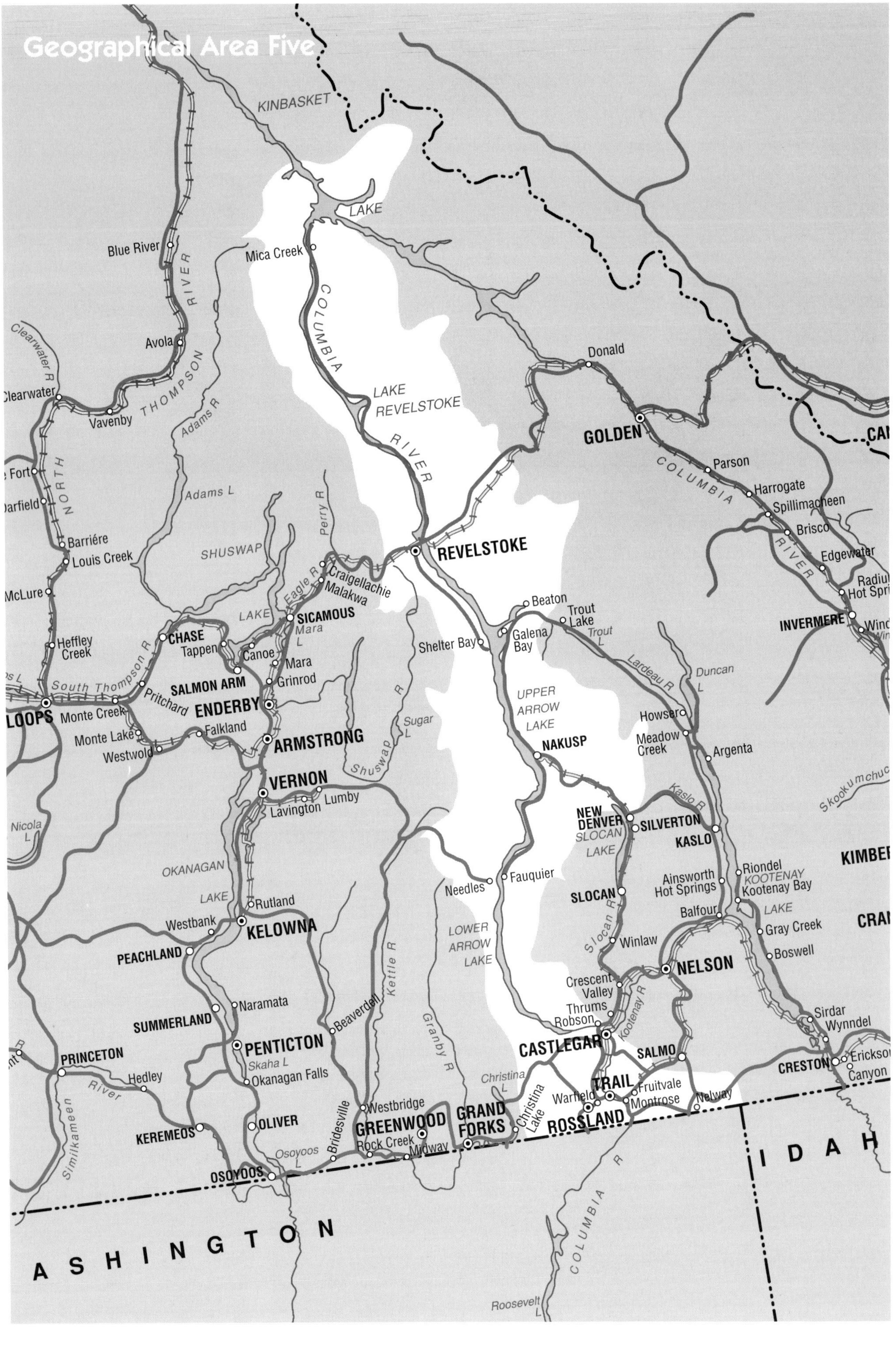
Geographical Area Five
KINBASKET
LAKE
Mica Creek
COLUMBIA
LAKE REVELSTOKE
RIVER
Donald
GOLDEN
COLUMBIA
RIVER
Parson
Harrogate
Spillimacheen
Brisco
Edgewater
INVERMERE
Blue River
NORTH THOMPSON RIVER
Clearwater R
Avola
Clearwater
Vavenby
Adams R
Adams L
Barriére
Louis Creek
McLure
Heffley Creek
SHUSWAP
LAKE
Perry R
Eagle R
Craigellachie
Malakwa
REVELSTOKE
Beaton
Trout Lake
Trout L
Galena Bay
Shelter Bay
Lardeau R
Duncan L
SICAMOUS
Mara L
Mara
CHASE
Tappen
Canoe
Grinrod
SALMON ARM
South Thompson R
Pritchard
Monte Creek
Monte Lake
Westwold
Falkland
ENDERBY
ARMSTRONG
Sugar L
Shuswap R
UPPER ARROW LAKE
NAKUSP
Howser
Meadow Creek
Argenta
Kaslo R
VERNON
Lavington
Lumby
Nicola L
OKANAGAN LAKE
Rutland
Westbank
KELOWNA
PEACHLAND
Naramata
SUMMERLAND
PENTICTON
Skaha L
Okanagan Falls
Beaverdell
Kettle R
Granby R
PRINCETON
Hedley
River
Similkameen
KEREMEOS
OLIVER
Bridesville
Osoyoos L
OSOYOOS
Westbridge
GREENWOOD
Rock Creek
Midway
GRAND FORKS
Christina L
Christina Lake
NEW DENVER
SLOCAN LAKE
SILVERTON
KASLO
Needles
Fauquier
LOWER ARROW LAKE
SLOCAN
Slocan R
Winlaw
Ainsworth Hot Springs
Riondel
KOOTENAY LAKE
Kootenay Bay
Balfour
Gray Creek
Boswell
NELSON
Crescent Valley
Thrums
Robson
Kootenay R
CASTLEGAR
SALMO
TRAIL
Warfield
Fruitvale
Montrose
Nelway
ROSSLAND
Sirdar
Wynndel
CRESTON
Canyon
Skookumchuck
COLUMBIA R
Roosevelt L

Geographical Area Five
Arrow Lakes/Revelstoke

Edgewood Lumber Company (later) William Waldie and Sons

WILLIAM WALDIE (SR.), owner of the EDGEWOOD LUMBER COMPANY on the Arrow Lakes in the early 1900s, shipped lumber by barge to Robson, where it was loaded onto rail cars for transportation to market.

Edgewood Lumber Company. Castlegar. Early 1900s.

William Waldie & Sons. Early 1940s.

William Waldie & Sons Ltd. Castlegar, B.C. Flood year, 1948. (Photo by Pat Archibald, Trail, B.C.)

In 1910, Waldie constructed a new mill at Castlegar and built an 85-foot steam tug to pull booms down the Arrow Lakes to the mill. Called The Elco, the tug's boilers were fueled with the sawmill slabwood.

Waldie's mill cut 60,000 board feet per shift and had its own shingle machine and dry kiln. In days when the interiors of homes were finished with lath and plaster, the mill produced lath as a by-product.

In 1928 the firm was named William Waldie and Sons. After William Sr.'s death in 1932, William (Bill) Waldie Jr. took over the company and kept it running through some lean years. The company replaced its horse-logging teams with tractor skidding in 1940. Business picked up during and after World War Two.

The Waldie's log supply came from wide ranging points in the Arrow Lakes basin. Wood-moving techniques were tailored to the region's diverse conditions. A five-mile water or "wet" log flume floated logs from Whatshan Lake to Arrow Lake. Log chutes and log drives were also employed.

Throughout his working life, Bill Waldie Jr. was a staunch supporter of ILMA. He sold the family operation to Celanese Corporation of America in 1952.

Celanese Corporation of America

After buying the Waldie interests in 1952, this company picked up other milling operations in the Arrow Lakes district. These included Columbia River Timbers at Sidmouth and Big Bend Lumber Company at Nakusp. Dennis Brookes, one of the Celanese management team, was ILMA President in 1966-67.

Celgar – First Pulp Mill in the B.C. Interior

The only pulp mill in the southern interior that ever had a "wood room" for chipping round logs, Celgar Ltd., part of the Celanese Corp., began making pulp in Castlegar in November of 1960. Much of the timber resource for the operation came from clearing the reservoir for the High Arrow dam (now the Keenleyside) at Castlegar. The reservoir flooded 80 miles of valley bottom in the Arrow Lakes and Columbia River Basin, from Castlegar to Revelstoke. The dam's design included a lock for the passage of log booms.

The Celgar pulp mill changed hands with successive changes in forest policy introduced by B.C. governments. In 1973, Premier Dave Barrett's NDP government bought 81 percent of Canadian Cellulose Co. from Celanese Corp. of America, which included the Celgar operations. The NDP government acquired Kootenay Forest Products Ltd. at the same time.

In 1979, Premier Bill Bennett's Social Credit government established the British Columbia Resources Investment Corporation (BRIC). This company included 81 percent of Canadian Cellulose, 100 percent of Kootenay Forest Products Ltd. at Nelson, 100 percent of Plateau Mills Ltd. near Vanderhoof and 10 percent of WestCoast Transmission Ltd.

In 1980 BRIC acquired the remaining 19 percent of Canadian Cellulose and merged all the forest companies under its control into B.C. Timber Ltd. B.C. Timber was later renamed Westar Timber Limited.

In 1986, the government sold the Celgar pulp operations to a consortium of interests from Montreal and Mainland China. Celgar closed its wood room in 1988, since its total chip requirement is now filled by sawmill waste from area mills.

Beginning in 1991 the pulp mill underwent a massive rebuild to bring it in line with new technology and environmental protection standards.

Hearn Brothers Lumber Co. Ltd.

Four brothers, Cliff, Willie, Howie and Eddie Hearn opened a mill at Nelway near Salmo in 1942. Following several expansion moves, they established a permanent sawmill and planer facility at Park Siding with the purchase of Hilco Logging. They sold out to Atco Lumber Ltd. of Fruitvale in 1976. Cliff's son Richie, continues on as Atco's mill superintendent.

Atco Lumber Ltd.
The Nelson Family

The Nelson family's history in logging has travelled full circle. ATLE NELSON SR. entered the industry in 1924. With a brother-in-law, he got involved with a Fruitvale sawmill in 1945. Atle then went into business with three of his sons, GILBERT, ATLE JR. (ADDY) and DICK to purchase F.A. VARSEVELD LTD., a holding that included a mill and private forest land. In 1959, they purchased the Fruitvale sawmill of NELSON LUMBER CO. LTD. This was the same sawmill that Atle Sr. had originally partnered with his brother-in-law in 1945. They incorporated in 1963 as ATCO LUMBER LTD.

Gilbert Nelson. Atco Lumber Co. 1966.

Over the next few years Atco Lumber bought the B.S. & L. LUMBER CO.'s forest land near Salmo, SULLIVAN CREEK FOREST PRODUCTS LTD. IN 1970, HEARN BROTHERS LUMBER CO. LTD. IN 1976, HADIKIN BROTHERS LUMBER CO. LTD. IN 1981 and GLENMERRY SAWMILLS in 1986. The company constructed a veneer plant in 1973. Fire destroyed the Park Siding mill (Hearn Brothers) in 1977. Atco rebuilt the Park Siding mill to modern standards the following year and it continues to operate. In 1984 the old Fruitvale mill was also destroyed by fire but not rebuilt.

Although the Nelson family has been the continuous owner/operator of Atco Lumber, many Nelson faces have changed. In 1973, Atle Sr. retired and sold his shares of the company to his two youngest sons, ALBERT and THEODORE (TED). Gilbert Nelson died in 1976, Atle Jr. (Addy) and Atle Sr. in 1990, and Dick in 1992. Albert and Ted remain as owners and managers.

Atco has been a member of the ILMA since 1961. Dick Nelson is well remembered as a long-term Board member and dedicated supporter of the ILMA.

Jones Ties and Poles (1978) Ltd.
Jack Bryden

JONES TIES AND POLES, located three miles south of Rossland, was started in the late 1940s by RUSSELL JONES, now deceased. Jones cut timbers and lumber for the local mines and the Trail smelter. JACK BRYDEN, the current owner, bought the company in 1978.

The operation continues making lumber for mines and the smelter but has now branched out into quality cedar and white pine cants for the U.S. market. In 1991 JONES TIES AND POLES put in a kiln and planer. In addition to mainstay products, the company now cuts furniture lumber, mostly white pine, for buyers in Vancouver. The mill employs 19 people.

D.B. Merry Lumber Co. Ltd.
Slocan City

D.B. MERRY started sawmilling in the Slocan City-Trail area in the 1920s. He moved to South Slocan and later to Casino, near Trail, and incorporated D.B. MERRY LUMBER CO. LTD. in 1937. The company established a retail building supplies operation at Trail at about the same time. In the late 1940s, milling operations were moved to Sheep Creek, near Rossland. The firm also operated in the Pend D'Orielle River area during this era. In 1960, the company name was changed to GLENMERRY SAWMILLS LTD. and the interest in the building supplies business was sold that same year.

D.B. Merry established a stationary mill complex in Trail in the early 60s and operated there until the mill was destroyed by fire in 1974. The mill was not rebuilt. D.B.'s son, GEORGE, operated in logging only from 1974. In 1986 the company's interest was sold to Atco Lumber Ltd.

D.B. Merry died in 1981. George Merry is retired in Trail.

Big Bend Lumber Co.
Sid Leary

A lumberman-turned-politician, Sid Leary started his company in Nakusp in 1930. He was elected MLA for the riding and also served as Minister of Highways in the 30s.

In the Lardeau River valley, at the north end of Kootenay Lake, an old wooden trestle bridge at Marblehead was known locally as "Leary's Bridge". Legend had it that the bridge was a make-work project of the depression years and that its real purpose was to curry favor with local voters by having government money spent in Leary's riding. You could drive over this excellent bridge, but once you did, you had to turn around and come back, because for years after the bridge was built there was no road on the other side.

After his death in 1946, Leary's widow managed the mill for four years before selling out to Wick Gray in 1950. The mill was sold to Celanese Corporation in 1952. Celanese shut it down and shifted the log supply to their big new mill at Castlegar.

Big Bend Lumber Co. 1953.

Marc Dumont and Sons
Slocan Lake

Marc Dumont and Sons was an early mill, established in 1936 at Hunter's Siding on Slocan Lake. It operated until 1950 when timber shortage forced a move. The Dumonts built at another setting, at Wilson Lake between New Denver and Nakusp. They set up a planer at a place called Brouse on the CPR branch line. Son, Paul, was in charge of logging, and another son, Carl, ran the sawmill. Dad took care of planing and shipping.

Dumont & Sons. Wilson Lake. 1951.

They sold to McDiarmid Brothers in 1960. Marc retired and the sons concentrated their efforts on another operation they had in Terrace. In 1964, they sold the Terrace mill and ran the Hoover Sawmill near Vernon until it was sold to Lavington Planer Mill in 1972.

Purcell Lumber Co. Ltd.
Bert Herridge and Erling Johnson

This little company was started in the early 50s in Nakusp by Bert Herridge and Erling Johnson. "Bert" Herridge also had a long career in federal politics as the CCF (later NDP) member for Kootenay West riding for more than 20 years.

Purcell Lumber Co. Ltd. was sold to Ian and Harvey McDiarmid in 1956 and Dave McInnes took over as manager. After they bought the Dumont operation in 1960, the McDiarmids sold all their interests in the Nakusp area to Axel Erickson of Passmore Lumber Co.

Dave McInnes moved to Vernon where he continued to manage forestry and logging for the mills under the McDiarmid Brothers and Ken Johnson.

Sine and Hess
Pioneer Mill Operators at Revelstoke

In 1903, Samuel Bowman of Minneapolis formed the Bowman Lumber Company and Revelstoke Lumber Company and took over three sawmills in the Revelstoke area. FRED SINE and FRED HESS, also from Minneapolis, took over as managers. One mill was destroyed by fire in 1910, resulting in relocation of one of the partners to Calgary. The other mills burned the next year. The partners' involvement with the lumber business continued in Alberta. They eventually acquired 14 lumber yards in Alberta, plus the one in Revelstoke.

In 1948 Sine and Hess bought KAMLOOPS LUMBER COMPANY and hired KEN LONG to manage their mill for them. Fred Sine died that year and Hess became president of their various companies. The firm went into the Redi-Mix cement business and when Fred Hess died, Sine's eldest son, BRETT, took over management of the business.

The firm continued to diversify with operations in Alberta. REVELSTOKE BUILDING MATERIALS LTD. emerged from these beginnings, eventually ending up as a Canadian owned company with 78 lumber yards and five concrete plants, as well as other ventures. In 1970, Revelstoke sold its mill operation in Kamloops to Weyerhaeuser Canada Ltd. It used most of the proceeds to expand its mill at Radium, purchased from the Wilder Brothers. That mill was sold to Slocan Forest Products Ltd. in 1985.

Downie Street Sawmills Ltd.
Vic Camozzi

VIC CAMOZZI was born in Revelstoke in 1908. In 1945 he started DOWNIE STREET SAWMILLS, so-named because of its location on a public thoroughfare called Downie Street. The mill was moved to a different location in 1949.

SAM OLYNYK became a partner in 1950 and STEVE MCKENZIE was taken into the partnership in 1955. In 1970 the operation was sold to KURT SHELEITH and GARTH LANGFORD who ran it until selling in 1979 to FEDERATED CO-OPERATIVES LIMITED. In 1988 it was bought by NELSON SKALBANIA, BILL NIMILOWICZ and JACK HEAVENOR. In 1990 the ROSS GORMAN family bought out Skalbania and Nimilowicz.

Recently, Downie Street Sawmills along with JOE KOZEK SAWMILLS LTD. and CASCADE CEDAR, assisted the City of Revelstoke in its purchase of a portion of the former Westar tree farm licence (TFL 23) and the formation of a municipal TFL. In return, these companies receive a percentage of the timber logged in the TFL.

Downie Street Sawmills has recently been renamed DOWNIE TIMBER LTD.

Chardon Lumber Co. Ltd.
Charlie Nakamura

CHARLIE NAKAMURA'S mill, CHARDON LUMBER CO. LTD., operated for an unknown period of time downstream on the Columbia River from Revelstoke. It went broke and was taken over by ED COOKE in 1962. Cooke renamed it COOKE-CHARDON LUMBER CO. Cooke's company also had a short lifespan as the mill was located on ground required by B.C. Hydro for the reservoir of the High Arrow dam at Castlegar. Hydro bought the property in 1965 and the operation passed into history.

Joe Kozek Sawmills Ltd.
Joe Kozek and Son Joe Jr.

For money owed him at the time, JOE KOZEK took over a portable sawmill in 1955. With hard work he did well enough to be able to move up to a larger operation at Revelstoke. He built a new modern mill at Revelstoke in 1972. In 1980, JOE KOZEK SAWMILLS LTD. put in a planer acquired from the Hadikin mill near Brilliant which had shut down.

At the death of JOE SR. in 1982, JOE JR. took over. In his 20s, he had already worked in the business for some years. The mill remains operating today and along with Downie Timber Ltd. is one of two working mills in the Revelstoke area. At one time there were more than 20.

Logging the Spruce Stands of the Revelstoke area, 1920's

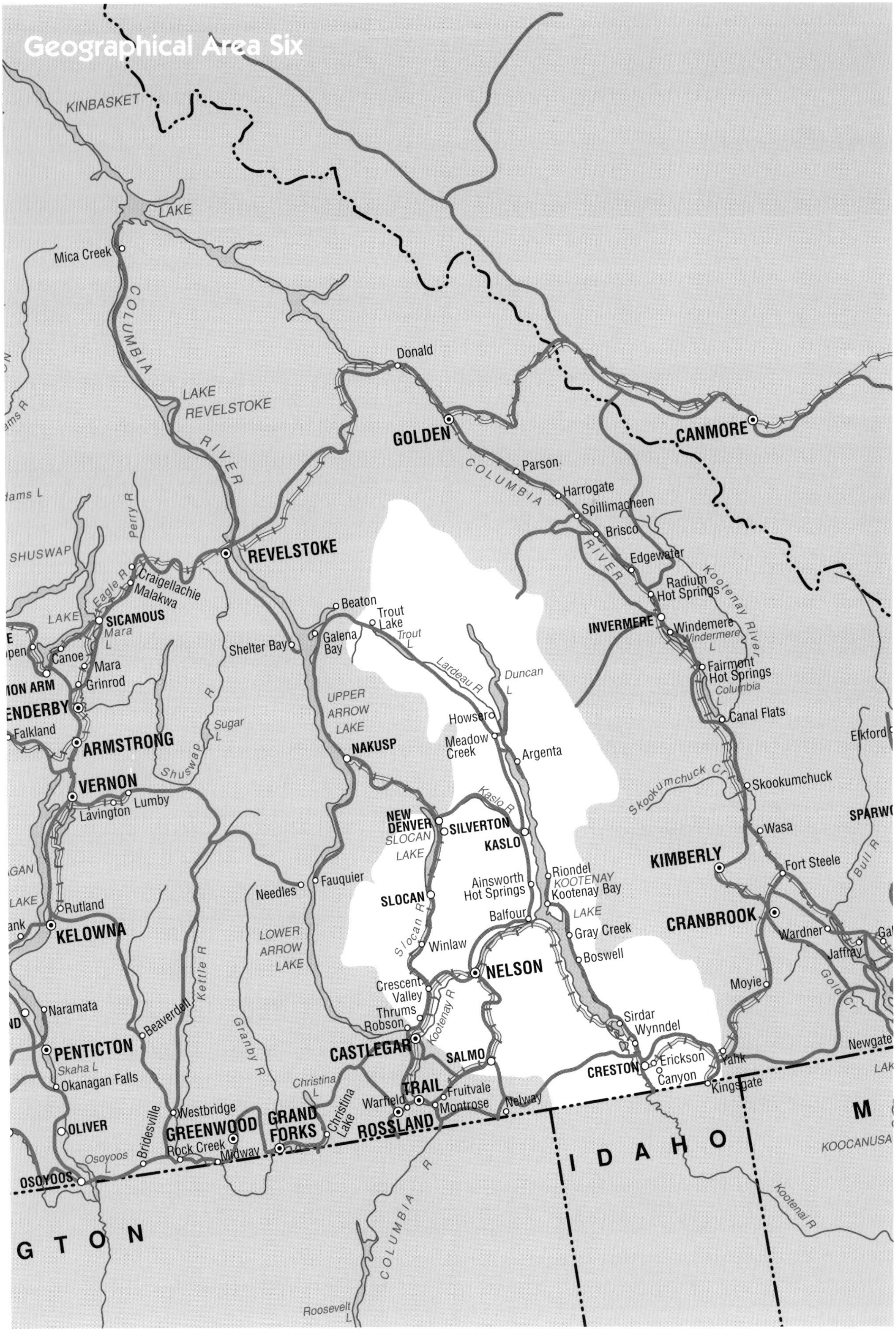

Geographical Area Six
KINBASKET
LAKE
Mica Creek
COLUMBIA
LAKE
REVELSTOKE
RIVER
Donald
GOLDEN
CANMORE
Parson
COLUMBIA
Harrogate
Spillimacheen
Brisco
RIVER
Edgewater
Radium
Hot Springs
Kootenay River
INVERMERE
Windemere
Windermere
L
Fairmont
Hot Springs
Columbia
L
Canal Flats
Elkford
Skookumchuck Cr
Skookumchuck
Wasa
SPARWOOD
Bull R
KIMBERLY
Fort Steele
CRANBROOK
Wardner
Jaffray
Gold Cr
Moyie
Newgate
Yahk
Kingsgate
KOOCANUSA
SHUSWAP
Perry R
REVELSTOKE
Eagle R
Craigellachie
Malakwa
LAKE
SICAMOUS
Mara
L
Canoe
Mara
Grinrod
ARMSTRONG
Falkland
Shuswap R
Sugar
L
VERNON
Lumby
Lavington
Rutland
KELOWNA
Naramata
PENTICTON
Skaha L
Okanagan Falls
OLIVER
Osoyoos
L
OSOYOOS
Bridesville
Rock Creek
Midway
Beaverdell
Kettle R
Westbridge
GREENWOOD
GRAND
FORKS
Granby R
Christina
L
Christina
Lake
ROSSLAND
Beaton
Trout
Lake
Trout
L
Galena
Bay
Shelter Bay
Lardeau R
Duncan
L
UPPER
ARROW
LAKE
Howser
Meadow
Creek
Argenta
NAKUSP
Kaslo R
NEW
DENVER
SILVERTON
SLOCAN
LAKE
KASLO
Riondel
KOOTENAY
Kootenay Bay
Ainsworth
Hot Springs
Needles
Fauquier
SLOCAN
Slocan R
Balfour
LAKE
Gray Creek
LOWER
ARROW
LAKE
Winlaw
Boswell
NELSON
Crescent
Valley
Kootenay R
Thrums
Robson
CASTLEGAR
SALMO
Sirdar
Wynndel
CRESTON
Erickson
Canyon
TRAIL
Fruitvale
Warfield
Montrose
Nelway
IDAHO
COLUMBIA
R
Kootenai R
Roosevelt
L

Geographical Area Six
Nelson/Creston

Rotter Sawmill
Frank Rotter

A sawyer who moved north to British Columbia from Washington State in 1926, Frank Rotter logged for several years before purchasing his own portable mill from the Christian Community of Doukhobours. In 1937 he started sawmilling in Sheep Creek south of Salmo.

By 1945 Rotter Sawmill was operating at peak production, exporting lumber to markets in the eastern U.S. Rotter sold out to Idaho Forest Products Ltd. in 1974 and died the following year.

Prime spruce logs bound for a pulp mill in Spokane via Great Northern Railroad. Rotter Lumber Co. 1939. (Photo by W.F. Montgomery)

Frank Rotter Sawmill near Salmo – 1939. (Photo by W.F. Montgomery)

Frank Rotter Sawmill near Salmo – 1939. (Photo by W.F. Montgomery)

Frank Rotter Sawmill near Salmo – 1939. (Photo by W.F. Montgomery)

Kootenay Valley Sawmills Inc. and Darkwoods

KOOTENAY VALLEY SAWMILLS was established in 1988 as a joint venture of VAAGEN BROTHERS LUMBER, Colville, Washington and PLUTO DARKWOODS CORPORATION, owned by the Duke of Wurttemberg in Germany. The company's site, 32 kilometers south of Nelson, was also used by LOUISIANA PACIFIC for loading chips onto rail cars on the Burlington Northern line. In 1991 Pluto Darkwoods bought the shares of Vaagen Brothers.

DARKWOODS FORESTRY, a subsidiary of Pluto Darkwoods Corporation, is unique among southern interior mills as the owner of a large tract of forest land. The company logs 56,000 hectares (140,000 acres) of private land in the Selkirk mountains between the south and west arms of Kootenay Lake. It is one of the largest private forest holdings in B.C. The property was originally a railroad grant – one of many blocks of land given to the CPR in return for building the first rail lines into the region.

Kootenay Forest Products Ltd.

KOOTENAY FOREST PRODUCTS LTD. has been known by several different names over its period of operation on Kootenay Lake. The mill was located in a bay at the north end of the City of Nelson. CADY LUMBER AND POLE COMPANY cut white pine lumber for the Nelson matchblock factory, W.W. POWELL. During the war, Cady sold out to PATTERSON LUMBER CO. and then it was bought by GLACIER LUMBER in 1946. Glacier was a subsidiary of DEER PARK LUMBER CO. LTD. of Spokane, Washington. Throughout decades of operation, the mill received its main log supply in booms from various logging sites around the Kootenay Lake basin.

Glacier built a new and modern mill on the site in 1950. It was sold to the EDDY MATCH COMPANY in 1952 and at that time was renamed KOOTENAY FOREST PRODUCTS LTD. (KFP). This new mill boasted a laminating plant which manufactured doors, beams and other finished products, as well as paneling and later, plywood.

KFP had planned to add a small log line to take advantage of close utilization and third band opportunities, but economics forced the plans to be shelved.

In 1973 the operation became part of CANADIAN CELLULOSE. In 1979 it was included in the list of operations of B.C. RESOURCES INVESTMENT CORPORATION (BRIC). The forest companies in BRIC became known as B.C. TIMBER LTD., later called WESTAR TIMBER LTD. The mill closed down in 1984 and the quota was broken up among other operators in the Kootenay Lake basin.

KFP was an outstanding contributor to the ILMA and VERN VANCE, one of its senior management people, served as ILMA President in 1958-59.

> *"I had been psyched to believe that the railroad was the only place to work so when I was laid off in the fall of 1959 I thought I had reached the end of the road. Getting a job as millwright at Kootenay Forest Products in Nelson was the beginning of considerable change in my life. One thing I noticed right away was that the Interior forest industry always seemed to be treated like second cousins to the big boys on the coast. They used to call us Jackpine Savages. Interior workers got paid less than workers on the coast and it didn't make sense to me that the spokesman for southern Interior workers was always a guy from Vancouver."*
>
> – JACK MUNRO, (RETIRED PRESIDENT, IWA CANADA)

Kootenay Forest Products Mill at Nelson – 1952.
(Photo by Stevens Studio Photographers, Nelson, B.C.)

Kalesnikoff Lumber Co. Ltd.

Kalesnikoff family, mid-1940s. Left to Right: Pete W. Kalesnikoff, Larry Postnikoff, Sam S. Kalesnikoff, Sam W. Kalesnikoff, Koozma Kalesnikoff, Mike Kalesnikoff (in foreground), John Kalesnikoff, Peter Kalesnikoff.

Even people with no connection to the forest industry marvel at the near pristine neatness of the Kalesnikoff mill yard on Highway 3-A at Thrums.

Since the KALESNIKOFF BROTHERS started their first mill in 1939, they have had their share of lean years. Today, their mill is a modern operation with a loyal crew and a reputation for conducting a well-managed operation, both in the bush and at the millsite. PETER Kalesnikoff remains active after 50 years in the business, and his son, KEN is now company vice-president.

Kalesnikoff Lumber was a hand/horse logging venture at the start. From their first setting in China Creek on the west side of the Columbia River, the family sold rough lumber to Waldie and Sons at Castlegar. Their first lumber truck was a 1937 Ford flatdeck. Conditions had improved by 1948, and they were able to acquire a new International TD9 tractor to build roads and replace horse skidding.

Kalesnikoff Lumber moved from China Creek to Robson Mountain in 1953. At their new mill site the power source was a TD18 diesel. Close to the time Waldie Lumber sold to the Celanese Corp., the Kalesnikoffs began to look for another buyer for their lumber. Passmore Lumber Co. took some of it and they also sold to Kootenay Forest Products Ltd. at Nelson.

In 1962, they moved to Sentinel Mountain and put in another mill. A forest fire nearly burned them out in 1967 and they did lose some machinery but the sawmill and planer survived. A heavy loss was a stockpile of over 3000 peeled cedar poles, decked and ready for shipment.

"We sent the crew up there but when the fire took off they had to run for their lives. I had several relatives working up there and they could all have lost their lives. My uncle knew the area and he took them to a side area because you could not stay ahead of the fire. We lost all the equipment and 3,000 to 4,000 cedar poles that were all ready to be picked up. We took a big chance to save the mill because we were surrounded by fire except for the very bottom. If the fire had turned and come in that way I think we would all be dead now. Thank God it went the other way around and we saved the mill and all our lives and we are still running the business."

– PETER KALESNIKOFF, (PRESIDENT, KALESNIKOFF LUMBER CO. LTD.)

In 1970 they leased and later bought property from Cominco at Thrums, B.C., the site of the present mill. Currently, this well-managed and well-run mill has dry kilns, barker and chipper facilities and two log lines, one for large wood and a second for small diameter logs.

In the environmental debate which has touched the entire forest sector, Kalesnikoff mill and bush operations have often been held up as a standard of responsible compromise between care for the land and efficient, competitive forestry.

Passmore Lumber Company Ltd. and Burns Lumber Co.

PASSMORE LUMBER was incorporated in 1938 to supply product for Burns Lumber Co., a wholesale and retail building supply firm in Nelson. GORDON BURNS, son of the founder was a forestry graduate from University of Washington.

Passmore Lumber built a new mill at Passmore (near Slocan) in 1945, and in 1950 was awarded the interior's first Tree Farm License.

"I am a great believer in Tree Farm Licenses. I worked for Passmore Lumber Company and it had TFL No. 3 which was the first in the interior of B.C. and was awarded in 1950. As a forestry student, I collected the data for this TFL. Even Peter Pearse, who did the last Royal Commission into forestry, held up the TFL as a model for good forestry practices and integrated use."

– JOHN MURRAY, (RETIRED VICE PRSIDENT, WOODLANDS DIV., C.F.I.)

Passmore Lumber Co. – 1950.

The operation was sold to AXEL ERICKSON of Colville, Washington in 1959. This era of consolidation in the interior lumber industry was marked by fierce competition between companies. Erickson planned to pick up all the mills in the Slocan and Nakusp Public Working Circles. PURCELL LUMBER of Nakusp, owned by IAN and HARVEY MCDIARMID (DAVE MCINNES was manager) had similar plans, and rumours flew concerning these competitive negotiations.

After a dramatic and tense period, with offers and counter-offers flying, finally AXEL ERICKSON and JOHN MURRAY (who recently retired as vice-president of Woodlands for Crestbrook) closed a deal with ROSEBERY LUMBER, near New Denver. This deal effectively killed Purcell Lumber's plan to expand south through the Slocan.

Eventually, Purcell Lumber sold to Passmore Lumber and following even more consolidation through the region, Passmore Lumber was sold to PACIFIC LOGGING LTD. of Victoria.

Pacific Logging sold to TRIANGLE PACIFIC FOREST PRODUCTS LTD. in 1970, and in 1977 these operations were sold to SLOCAN FOREST PRODUCTS LTD. under the control of IKE BARBER.

"In the early 1950s Burns Lumber Company had three camps with about 10 or 15 horses in each camp. Later we phased out the horses and went into jammers. These were something like a cable high-lead system only much smaller. Then we went into rubber tired skidders and light flotation

machines like the KMC skidder, made in the Kootenays, which literally float along the ground. And there were small crawlers and we also did some skyline logging with the mainline slung on intermediate supports. Then before I left Crestbrook we did three months of helicopter logging. So I've seen pretty well all the changes."

– JOHN MURRAY, (RETIRED VICE PRESIDENT, WOODLANDS DIV., C.F.I.)

Slocan Group
Ike Barber

Formed in 1978, the Slocan Group was known as SLOCAN FOREST PRODUCTS LTD. until 1994. Its original purchase was TRIANGLE PACIFIC FOREST PRODUCTS LTD. at Slocan. In 1979, the company purchased TRI-PAC'S Quesnel operations. In 1985, Slocan acquired REVELSTOKE CONCRETE AND LUMBER LTD., at Radium Hot Springs. In 1986 the company went public and acquired CLEARWATER TIMBER PRODUCTS LTD. Clearwater Timber Products Ltd. had mills at Clearwater, Vavenby, and Valemount in the North Thompson valley. Those operations were consolidated at Vavenby and Valemount. Slocan also purchased the assets of WEIR'S SAWMILL LTD. at Quesnel in 1987, TACKAMA FOREST PRODUCTS at Fort Nelson in 1989, FIBRECO PULP in 1991 and PLATEAU MILLS (from Westar) in 1992. Slocan Group and DONOHUE INC. are currently finalizing arrangements for the purchase of FINLAY FOREST INDUSTRIES LTD. in Mackenzie, B.C.

TERRY UPGAARD, vice-president Sales & Marketing, has been a Director on the ILMA Board since 1989.

"...and you know, when you talk to the bank they will loan you some money as long as you have some money to put up. We were a group of boys that had no money. My dear wife who had title to our house because we had burned (paid off) the mortgage a few years back, re-mortgaged the house. As she gave me the money from the house she said; "I hope you guys know what you're doing." That got us $80,000 which gave us enough that we could borrow the rest and that's how we got started."

– IKE BARBER, (PRESIDENT, CHAIRMAN & CEO, SLOCAN GROUP)

Fairhurst Sawmill
The Fairhurst Family

Incorporated at Silverton in 1948, FAIRHURST SAWMILL operated in the Slocan Public Working Circle until selling out to Passmore Lumber in 1960. BOB FAIRHURST was company president and general manager and MARGERY FAIRHURST was secretary.

Rosebery Lumber Co. Ltd.
Jordan and Yamada

Local portable mills supplied a planer at Rosebery with rough lumber. The planer mill was opened by HOWELL JORDAN and ROY YAMADA in 1954. Main portable mill suppliers were owned by Roy Yamada and K. ONO in Wilson Creek near Nakusp, and by MURAKAMI BROTHERS at Hills, between Nakusp and New Denver.

Rosebery Lumber Co. Ltd. sold to Passmore Lumber in 1960.

Stafford and Sons
Kaslo and Procter

RAY STAFFORD began logging in Eugene, Oregon in 1929 and expanded into sawmilling. In 1957 he and his sons, BELDON and LYLE, moved to B.C. and set up a mill at Kaslo under the name of H.R. STAFFORD and Sons Ltd. In 1959 they moved to Procter on the West Arm of Kootenay Lake.

Logs were boomed to the mill from bush operations in the Lardeau Valley at Kootenay Lake's north end. The Procter mill, including dry kilns and planer, produced lumber for shipment to the prairie provinces. The Stafford operation was sold to Kootenay Forest Products in 1968 and the mill was closed down.

T and H Sawmills
Kurt Thomas and Bob Howlett

Steel salvaged from the old Granville Street bridge in Vancouver built the first T AND H SAWMILL. The bush mill was located near Kaslo, at a logging camp that had been used to accommodate and employ displaced Japanese Canadians during World War Two. The mill was built by KURT THOMAS, who was born and raised in the Kootenay Lake country. Thomas ventured into sawmilling with BUDDY WARD and, later another partner, BOB HOWLETT.

They purchased a planer from S. M. SIMPSON of Kelowna in 1959 and in 1967 built a new mill and booming ground on the beach in Kaslo. This mill was destroyed by fire and replaced with a new facility which continued operations until forced into bankruptcy by the recession of 1982. Bob Howlett lives in Calgary and Kurt Thomas still resides in Kaslo.

Meadow Creek Cedar Ltd.
Albert Volpatti

ED ALDINGER and CECIL PANGBURN JR., son of a pioneer logger, started Meadow Creek Cedar in 1976. Two years later, Pangburn left the operation and Aldinger's brother, Horst, joined as a partner. ALBERT VOLPATTI became a third partner in 1985. The company was purchased by TAIKI INDUSTRIES of Japan in 1989. Taiki built a new and modern facility at the millsite after a devastating fire in their first year of ownership.

The new mill has a payroll of about 90 and contractors employ another 25 to 30. The company quota is 118,000 cubic meters per year. The mill cuts approximately 154,000 cubic meters per year, obtaining volume in excess of Crown quota by log purchases. The company ships 95 percent of its lumber production to a distribution center in Japan.

Gray Creek Forest Products Ltd.
Oscar Wirsigg Family

This family operation, located at Gray Creek on the east side of Kootenay Lake between Crawford Bay and Creston, dates back to the 50s. It was sold to KOKANEE SPRINGS DEVELOPMENT LTD. in 1965 as part of a planned recreational development at Crawford Bay. After Kokanee Springs sold to CRESTBROOK FOREST INDUSTRIES LTD. in 1967 the operation was shut down.

Wynndel Box and Lumber Co. Ltd.
In the same family for 81 years

On property owned by his father, O.J. WIGEN, MONRAD WIGEN began in 1913 to make boxes and crates to supply the local tree fruit and berry industry. In Wynndel, just north of Creston, Monrad's steam-powered box factory operated during the summer months, supplied with wood by the same portable mill that cut railroad ties for the CPR during the winter months.

Wigen's company, incorporated in 1923 as WYNNDEL BOX & LUMBER CO. LTD., reduced its output through the depression years. In 1938-39 the company moved to the present site at Wynndel and built a new box factory, also powered by a steam engine. In 1945-46 a sawmill was constructed. Over the years, the operation has continued to modernize. Production went to specialty items such as pattern stock, siding, paneling, decking and other value-added items. Although the box business was eventually phased out, the company kept its original name.

> *"We are fortunate in that we are now in the fourth generation. The company was started by my father and grandfather. Michael and David are now the fourth generation. We are one of the small operations right now. Maybe the writing is on the wall. I hope not. We think there is still a place for smaller mills."*
>
> – JACK WIGEN, (PRESIDENT, WYNNDEL BOX AND LUMBER CO. LTD.)

Since Monrad's passing in 1971, the company has been operated by his sons, JACK and BOB, who have been joined by their respective sons, MICHAEL and DAVID, making a continuous family presence of four generations dating from the year before the First World War.

Jack Wigen is a long-serving member of the ILMA Board of Directors.

Creston Sawmills Ltd.

The town of Creston literally grew up around CRESTON SAWMILLS LTD., a company incorporated in 1904 by CHARLES O. ROGERS. Rogers' mill was taken over by DON BURNS and ALF FARSTAD in the 1940s. A veneer plant was opened in 1963 and the company sold out to CRESTBROOK TIMBER LTD. the next year. GORDON FISHER, Company Manager, was President of the ILMA in 1980-81. At its close in 1991, Creston Sawmills had been operating for 87 years within the city limits of Creston.

Alf Farstad (in hat) on old slab truck. 1932.

J.H. Huscroft Ltd.
Another long-time family operation

Just north of the Idaho border, a father and son started a tie mill on the family farm. In 1927, JOHN HENRY HUSCROFT and his son, ELMER delivered their first load of ties to the Great Northern Railroad at Porthill, Idaho, a two-mile distance, with a team of horses. In the next 25 years, the Huscroft family would become experts in the tie business.

John's son, Burt, died in a logging accident in 1942, and Elmer died, as a result of a sawmill accident, in 1953. Other members of the family continued and expanded the operation. A portable sawmill that burned in Dodge Creek in 1961 was not covered by insurance. The Huscroft's believed, as did many other operators at the time, that insurance rates for portable mills were too high to justify the expense. A year after the fire, the Huscrofts built a permanent mill at Erickson, just east of Creston.

Brothers KEN and LEONARD HUSCROFT continued the family tradition through the 1950s. Ken bought Leonard's interest in 1961 and is currently president of the company. His daughter, GWEN TELLING, of the third generation, is a senior manager of the company and serves as a Director on the ILMA Board.

"My Dad was steam engineer on the tugboat Hercules. They towed logs down the Kootenay River – some from Idaho – and they went to Nelson. That would have been between 1914 and 1920."

"My Dad and older brothers started the first sawmill in 1927. I am still there every day from morning to late but I am getting rather old and I am mechanical anyway so I don't spend much time at the office. I just stay down at the mill. My daughter Gwen has been running the show for the past 15 years now so it has gone from father to son to grandchildren. We also have a few good key men and that really helps. One of them lied to me once. He told me when I hired him that he was 18 years old. Later I found out he was only 16. That was about 40 years ago. He's still with me and it was the only time he ever lied to me."

"We shut our sawmill down in the winter of 1948/49 and put our machinery to work repairing dikes after the famous '48 flood. While we were away, our mill up Boundary Creek burned down. There was a large sawdust pile there and the fire probably started in that. The mill was destroyed, we had no insurance and the fire was still burning. Our mill was only 100 yards from Boundary Creek and another small creek

ran right past the mill. So I took a bulldozer and just diverted the creek through the sawdust pile to put the fire out. I washed all the sawdust into Boundary Creek and it went on downstream. If you were to do that today you would be put in jail for the rest of your life."

"To me the most important change for the better is safety. It seems that people are more safety conscious now than in previous years. We have had two fatal accidents in our family business. One was in 1942 and the other in 1953. Both were my brothers."

– Ken Huscroft, (President, J.H. Huscroft Ltd.)

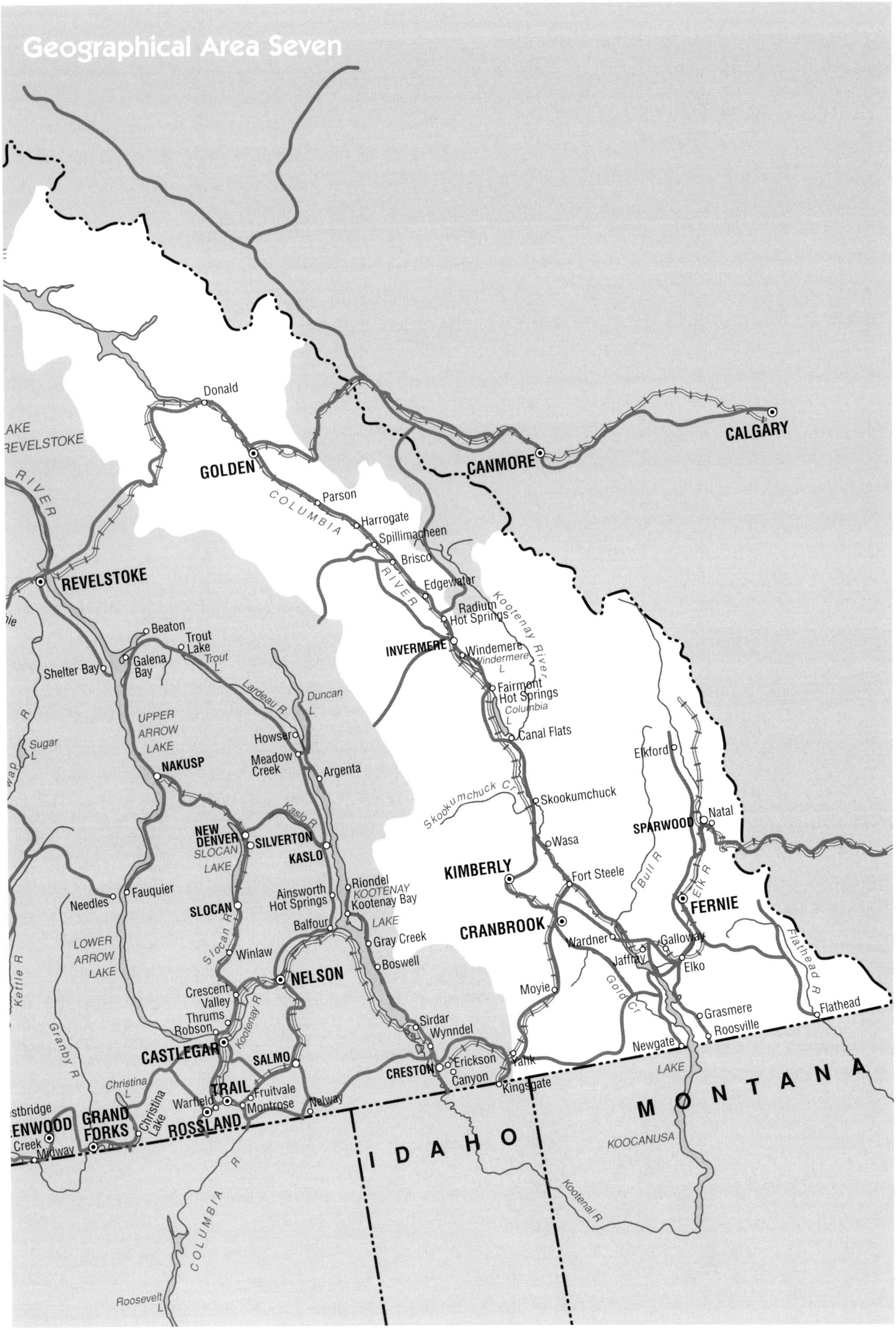
Geographical Area Seven
Donald
GOLDEN
CANMORE
CALGARY
COLUMBIA
Parson
Harrogate
Spillimacheen
Brisco
RIVER
Edgewater
Radium Hot Springs
Kootenay River
INVERMERE
Windemere
Windermere L
Fairmont Hot Springs
Columbia L
Canal Flats
REVELSTOKE
RIVER
Beaton
Trout Lake
Trout L
Galena Bay
Shelter Bay
Lardeau R
Duncan L
UPPER ARROW LAKE
Sugar L
Howser
Meadow Creek
Argenta
NAKUSP
Kaslo R
NEW DENVER
SILVERTON
SLOCAN LAKE
KASLO
Skookumchuck Cr
Skookumchuck
Wasa
Elkford
SPARWOOD
Natal
KIMBERLY
Fort Steele
Bull R
Elk R
FERNIE
Riondel
KOOTENAY
Kootenay Bay
LAKE
Ainsworth Hot Springs
Balfour
Fauquier
Needles
SLOCAN
Slocan R
LOWER ARROW LAKE
Kettle R
Winlaw
Gray Creek
Boswell
NELSON
CRANBROOK
Wardner
Jaffray
Galloway
Elko
Flathead R
Gold Cr
Moyie
Grasmere
Roosville
Flathead
Crescent Valley
Thrums
Robson
Kootenay R
Granby R
CASTLEGAR
SALMO
Sirdar
Wynndel
CRESTON
Erickson
Canyon
Yahk
Kingsgate
Newgate
LAKE
MONTANA
Christina L
Christina Lake
Warfield
TRAIL
Fruitvale
Montrose
Nelway
GRAND FORKS
ROSSLAND
Creek
Midway
IDAHO
KOOCANUSA
Kootenai R
COLUMBIA R
Roosevelt L

Geographical Area Seven
East Kootenay/Columbia

McDonald Ranch & Lumber Ltd.
Grasmere, B.C.

In 1923, JOHN A. MCDONALD settled just north of the Montana border, in the Grasmere area near Elko. From small beginnings, McDonald built a flourishing sawmill and cattle business that was eventually run by his two sons, DOUG and ANDY. A unique feature of the McDonald operation has been a reliable hydro-electric generator that has supplied an abundance of electric power.

A third generation is now active in the management of MCDONALD RANCH & LUMBER. Andy's son CAM, is involved in the agricultural side of the enterprise while Doug's sons BARRIE and GLENN are involved with the sawmill and woodlands activities. Doug McDonald served as a Director of the ILMA for a number of years.

Early logging truck at Grasmere, B.C. (near Montana border).

Grasmere, B.C. (near Montana border).

Yahk Lumber Co.

Ross – Saskatoon Co. camp, Waldo/Elko area.

Ross – Saskatoon Co., Waldo/Elko area.

E.C. Letcher Lumber Co.

In 1921 Tom and E.C. (Ted) Letcher went into the tie business at Grasmere. In the early 1940s they built a larger lumber mill near Grasmere which, in 1944, they moved to Gold Creek, west of Waldo and south of Elko. This mill was operated by Ted and his wife, May.

The Letchers sold to Crow's Nest Industries in 1965 and moved back to Grasmere. Doug and Andy McDonald (see last entry) are nephews of Ted (who passed away in July, 1991) and May Letcher.

Wilson Lumber Co. and East Kootenay Lumber Co.

James White worked as an accountant for the Wilson Lumber Company in the 1930s. In partnership with John Jacubiec, White took over the company in 1945 and changed the name to East Kootenay Lumber Co. Ltd. White served as ILMA Chairman in 1948-49 and as Mayor of Fernie from 1956 to 1964.

East Kootenay Lumber had portable sawmills in the Newgate and Cedar Valley areas, between Elko and Wardner, and a planer mill in Fernie. The mill was sold to Crow's Nest Pass Lumber shortly before White's death in 1964.

Crow's Nest Pass Coal Company
(later Crow's Nest Industries Limited)

To support its mining operations, CROWS NEST PASS COAL COMPANY operated logging camps and a sawmill, beginning in the last few years of the 19th Century. In 1950 the company bought a planer and went into the lumber business in a serious way. A new sawmill was built at Natal in 1959. By 1963, with its name changed to CROW'S NEST INDUSTRIES LTD. (CNI), the company began a series of acquisitions to increase its annual cut. It bought up ANCHOR LUMBER CO. that year and BODIO LUMBER CO., EAST KOOTENAY LUMBER CO. and REVELSTOKE LUMBER CO. in 1964. In 1965 CNI acquired E.C. LETCHER LUMBER CO., ST. MARY'S LUMBER CO. and FLATHEAD LUMBER CO. By the time these holdings were assembled, CNI had a cut of 68 million board feet per year from five stationary and three portable mills.

After CNI built a large modern mill at Elko in 1969, it closed its outlying mills. In 1983 CNI's operations were sold to CRESTBROOK FOREST INDUSTRIES.

Knight Lumber Co. Ltd. and Gold Creek Timber

Three miles west of Crow's Nest, FRANK KNIGHT started a lumber mill in the early 1940s. ED CUMBERLAND and GUS ERIKSSON joined in partnership in 1944, and the operation expanded with portable mills in the area. In 1949 KNIGHT LUMBER CO. LTD. sold out to Cumberland and Eriksson.

With other companies in the region, Knight Lumber formed GOLD CREEK TIMBER COMPANY LTD. Gold Creek's purpose was to monitor timber sale auctions in the region to try to cool down the destructive bidding wars that were common in that period. The Gold Creek alliance was established to address the same problems that led to the formation of K-V TIMBER LTD., which was later incorporated in the Boundary country.

In 1960, Eriksson and Cumberland sold Knight Lumber to Revelstoke Sawmills Ltd. Gus Eriksson was President of the ILMA in 1955-57.

Galloway Lumber Company Ltd. and MacDonald Lumber

In the late 1890s, MacDonald Lumber built a sawmill and planer on the current site of Galloway Lumber Co. Ltd. In 1929, the sawmill was purchased by MIKE DUMONT. Sixteen years later, in 1945, Dumont sold the business to OSCAR JOSTAD and CHARLIE NELSON. The company, renamed GALLOWAY LUMBER CO. LTD., was managed by HENRY NELSON, Charlie's son. Over the next few years, Galloway Lumber bought several smaller operations in the area including GRAF BROTHERS, SILVER RIDGE SAWMILLS and WIGWAM LUMBER. By 1970 Galloway had bought out the Jostad interests. The company has been owned and managed as a family business to the present.

Henry Nelson (now deceased) was President of the ILMA in 1972-1973. His son BUD is on the ILMA Board of Directors.

"Our company is a relatively small company and we do most of our business on a handshake. The only written contracts are between the government, ourselves and the banks. Otherwise the handshake seems to work all right. If anyone is going to go his own way he is going to do it regardless of whether there is a contract or not."

"When the quota system came in some of the companies decided they could get security of timber supply and they started buying up all the little companies. We are down to three companies in our area; one small, one medium, and one very large company as a result of that quota system. The quota has provided security to the industry and the workers. Lots of companies amalgamated to take advantage of the quota system."

– BUD NELSON, (PRESIDENT & GENERAL MANAGER, GALLOWAY LUMBER CO. LTD.)

Cranbrook Timber Products Ltd.
Tingstad Brothers

CRANBROOK TIMBER PRODUCTS LTD. operated from the late 1940s, until its sale to JSB TIMBER CO. LTD. in 1959.

JSB Timber and Fuel Products – an unusual beginning

Double and triple partnerships have been common in the sawmilling business. But by the time JSB TIMBER AND FUEL PRODUCTS had been in operation for just a short time, there were already 15 owners involved in the partnership.

JSB Timber started cutting ties near New Lake, outside Cranbrook, in 1942. It bought a second mill from ERIC MCKINNON of Cranbrook Foundry in 1945. The partners moved it to Lumberton, an historic "lumber town" that once existed between Creston and Cranbrook. They named the logging part of their operation JSB LUMBER CO. LTD. and built a planer near Cranbrook. Ten of the partners then sold out to remaining partners WALTER BARRETT, ED HORSMAN, BOB SPARKS, BERT NILSEN and JOHN KURRE. These five partners built a mill at Moyie in 1956. In 1959 they bought out CRANBROOK TIMBER to add quota to their holdings. In 1964 they sold their operations to CRESTBROOK.

Sunday – noon meal at Weatherheads Camp. Yahk, 1925.

Kennelly Lumber Co.

Two partners, with surnames BARR and KENNELLY, started cutting ties in the Cranbrook area in 1941. They formed KENNELLY LUMBER CO. in 1950 in the City of Cranbrook. They continued to operate until 1967 before selling out to CRESTBROOK.

Cranbrook Cartage and Transfer Co. Ltd.
Jostad and Nelson

During the First World War era, partners OSCAR JOSTAD and CHARLIE NELSON tried their hand at various enterprises in the Cranbrook area. Their company, CRANBROOK CARTAGE AND TRANSFER CO. LTD., purchased a small sawmill at Bull River, east of Cranbrook, in 1928. After trying to make a go of the lumber business, the partners closed their mill and went back into the cartage, livery and dairy businesses. In 1938 Jostad and Nelson decided to give the sawmill business another try. They set up portable sawmills at Gold Creek and at Fort Steele, as well as a planer in Cranbrook. Eventually they were bought out by the Nelson family of GALLOWAY LUMBER.

Several years' accumulation of planer shavings on the millsite were dozed flat after the mill closed. Topsoil was added and the site now serves as a playing field for the City of Cranbrook.

Leask & Johnson, Cranbrook, B.C.

Cranbrook Sawmills Ltd.

CRANBROOK SAWMILLS LTD., started by DON BURNS and ALF FARSTAD in the late 1940s, was acquired by Crestbrook in 1956.

Cranbrook Sash and Door Co. Ltd.

A four-way partnership between HARRY MCCOWAN, HARRY SPENCE, BILL ATRIDGE and JOE WOODMAN, gave rise to yet another East Kootenay tie-cutting operation. After cutting ties through the 1930s, the partners added a planer in 1941. Their Cranbrook planer was fed with rough lumber supplied by local portable mills, as well as by their own mill.

A background in banking qualified VIC BROWN as the young timber company's office manager. After 1947 he managed the company's affairs. Brown was Chairman of the ILMA in 1952-53. CRANBROOK SASH AND DOOR followed the trend of the times by buying up smaller mills in the region. The company was purchased by Crestbrook in 1956.

Silver Ridge Sawmills Ltd.

Silver Ridge Sawmills operated from several locations in the Cranbrook area through the 1950s and 60s. Among principal characters involved were founders Vince and Joe Downey, Steve Shypitka and Bill Griffiths. The mill was sold to Galloway Lumber in 1987.

"Packing in" (Lumberton area).

Lumberton log inventory.

Kootenay Wood Preservers Ltd.

MEL BILDFELL and BILL MATHER started this Cranbrook-based business in 1979. They expanded into Alberta in 1987, and added a pole operation in Brisco in 1990. KOOTENAY WOOD PRESERVERS currently employs 160 people in the manufacture of pressure treated lumber, fence posts, agricultural rails, utility poles and other value added products. Mel Bildfell served on the ILMA Board for five years in the mid-1980s.

Crestbrook Forest Industries Ltd.
A dramatic history of growth and consolidation

The seed that would eventually grow into a large forest products company was planted with the formation of SYLVAN INDUSTRIES LIMITED. SYLVAN was formed in 1955 to buy CRANBROOK SASH AND DOOR and CRESTON SAWMILLS. The dual acquisition took longer than expected, but was finally accomplished in 1964. The newly-formed company adopted the name CRESTBROOK TIMBER LTD. and soon bought out a number of smaller mills in the area. By the late 60s, the company operated four sawmills, a veneer plant at Creston, a pulp mill at Skookumchuck and a plywood plant at Fort MacLeod, Alberta.

The company's expansive history is highlighted chronologically:

1955 Crestbrook Timber Limited incorporated.
1956 Purchased Cranbrook Sash and Door.
Purchased Cranbrook Sawmills.
Purchased Christian Wenger Lumber (Skookumchuck).
1963 Purchased Colin Cameron Lumber (Wasa).
1964 Purchased JSB Lumber.
Purchased Creston Sawmills.
1967 Purchased Kennelly Lumber.
Purchased Gray Creek Forest Products.
1983 Purchased Crow's Nest Industries.

The pulp side of Crestbrook's operation was made possible with additional investments by HONSHU PAPER CO. LTD., and MITSUBISHI CORPORATION, both of Japan. Crestbrook was able to construct a $43 million, 400 ton-per-day bleached kraft pulp mill at Skookumchuck. In 1967 the company name was changed to CRESTBROOK FOREST INDUSTRIES LTD. (CFI).

Under the leadership of STUART LANG, who joined CFI as President in 1979, the company has continued its pattern of growth in B.C. and into Alberta. In 1988 Crestbrook and two Japanese investors received Government of Alberta approval for the phased development of a $1.3 billion pulp mill in the Athabasca – Lac La Biche area, 145 kilometers north/east of Edmonton. Construction of the "Alpac" mill began in 1991 following lengthy scrutiny through a provincial and federal environmental assessment process. Final government permit approvals were granted in 1993.

Crestbrook, a generous contributor to ILMA over the years, has made staff time available on a number of occasions. MIKE ROUSE, Vice-President, Wood Products for CFI, was ILMA Chairman in 1988-89 and is currently an ILMA Director.

"We have all become more aware of the importance of the many values in the forest. If we considered them before it was more by instinct than deliberate consideration. In the enhancement of wildlife, we are much more cognizant of the impacts of various activities."

"It was about 1980 when we realized that this environment concern was not just a passing fad. We realized that this was something that was going to concern every single person and it was going to continue to grow. Of course it has had a tremendous effect on our business."

"Thirty to 40 percent of the economy in the East Kootenay is dependent on the forest industry. That, along with the mining, hasn't changed much over the last few years. The only thing that's changed is that the fundamentals of local economies have become far more difficult for the average person to understand."

– MIKE ROUSE, (VICE PRESIDENT, WOOD PRODUCTS, C.F.I.)

Colin Cameron Lumber Co.

Colin Cameron, a forest ranger at Fort Steele in the early 30s, left the Forest Service to operate a portable sawmill at Fort Steele. In 1939 he moved to Wasa, north of Fort Steele, and constructed a stationary mill. After his death in 1963 following a sudden heart attack, the Cameron company assets were acquired by Crestbrook.

Colin Cameron with Fargo logging truck, Wasa, B.C. Spring of 1952

Columbia Contracting Co.
Pat McGrath and Dick Jarvis

In the early 1940s Pat McGrath and Dick Jarvis bought a mill from the CPR at Canal Flats. They named their company Columbia Contracting Co. Ltd. In 1950, the company was bought out by Cranbrook Sash and Door.

Christian Wenger Lumber Co. Ltd.

Swiss emigrant Christian Wenger arrived in Kootenay territory in 1937 and bought a farm near Invermere. Filling an order for 100,000 Christmas trees for Hoffert Ltd. provided the capital for Wenger to build a sawmill on the Westside Road near town. In 1945 he built 25 miles of road up the Skookumchuck River, built a logging camp and sawmill and launched into the lumber business as Christian Wenger Lumber Co. Ltd. He suffered a heart attack in 1956 and sold his business to Crestbrook. He died in Cranbrook in 1990 at age 93.

Simon Ronacher & Son Limited

Simon Ronacher, an emigrant from Salzburg, Austria, came to the Kootenays in 1927. He hauled ties for the CPR into the 1930s, and in 1935 built a planer mill at Athalmer, near Invermere. He planed rough lumber from local portables and expanded with two portable mills of his own in the late 1940s. After his death in 1954 his son, John, took over the business, built a stationary mill and centralized operations at Athalmer. The company continued in business until its 1967 sale to Revelstoke Building Materials Ltd.

Simon Ronacher and Jack Evans March 1947.

Ronacher mill Athalmere, B.C. 1955.

Seel Forest Products Ltd.

An independent, family-owned enterprise at Edgewater, near Radium Hot Springs, SEEL FOREST PRODUCTS LTD. was started in 1940 by HEINZ SEEL and a crew of six employees. Still in the Seel family, the business currently employs approximately 35 people. Many changes have taken place, from horse logging and portable milling to mechanized logging and a fully-equipped stationary mill which utilizes the latest in computerized systems. A remanufacturing plant has also been established, along with dry kilns.

Approximately 35% of current production is in timbers. The remaining 65% is in kiln dried dimension and metric sized lumber. This material is further processed for various purposes – such as for furniture blanks, window and door frames, tongue and groove decking, patterned stock, panelling and siding.

These products are marketed by several agents throughout Canada, United States, Japan and the United Kingdom.

Coulter and Warner Planer Ltd.
Sid Warner and W.R. Coulter

To take advantage of the rough lumber resource produced by numerous portable sawmills in the region, partners surnamed COULTER and WARNER opened a planer mill in the late 1950s. Warner sold out to Coulter, and in 1969 the company went into bankruptcy. It was then acquired by OJAR UKASS and FRITZ ZEHNDER, who renamed the company NORTH STAR PLANING CO. LTD. Various changes in partnership arrangements involved LEO BUKOVNIK, ALBIN SCHAGER and others. The company is now managed by LEO BUKOVNIK JR. North Star has diversified to include a retail center with a full line of building supplies.

"My father Leo Bukovnik Sr. came down to manage the company. He managed it for two years and then bought out Ojar Ukass and Fritz Zehnder who had been using the name of North Star Planing Co. Ltd. That

was in 1976. He took in my uncle, Albin Schager, as a partner and built the business up. I took over in 1986 and we started the building supply business about 1987. Now we do our own sawmilling, planing and logging and we are up to about 80 percent now in the retail market. My father retired in 1991 and I have been the general manager since then. What we are doing now is trying to retail as much as we can and sell off the excess in the wholesale market. We have our own loggers, our own equipment, our own sawmill, our own planer and our own retail business. We go from the stump to the cash register. We are our own middle man."

– Leo Bukovnik Jr., (General Manager, North Star Planing Co. Ltd.)

Wilder Brothers Lumber Co.
Lloyd and Earl Wilder

Brothers Lloyd and Earl Wilder had been in the trucking business in Alberta before moving to the East Kootenay in 1945. Their venture into the lumber business started small, with rented machinery and a portable tie mill. As Wilder Brothers Lumber Co., they built a stationary sawmill at Fairmont Hot Springs. In the early 1950s they relocated to a new sawmill and planer at Radium. They bought out L & Q (Lemaster and Quinn) Lumber at Canal Flats and acquired other small mills in order to add to their timber supply. They sold out to Revelstoke Sawmills in 1964.

Revelstoke Sawmills Ltd.

After purchasing the Radium sawmill from Wilder Brothers, Revelstoke Sawmills embarked on a major expansion. Aided by capital from the sale of its Kamloops mill to Weyerhaeuser, Revelstoke Sawmills purchased Edgewater Sawmills Ltd. and the mill at Athalmer owned by Simon Ronacher and Son Ltd. In 1985, the Radium mill was sold to Slocan Forest Products Ltd.

Stone and Gillis Sawmills Ltd.
Jim Stone and Dick Gillis
Also Brisco Sawmills Ltd., the Lautrup brothers

Supplied with rough lumber from a portable mill, partners Stone and Gillis started a planing operation at Brisco in 1950, and in 1962 they built a stationary mill beside the planer. They sold out to Ken and George Lautrup in 1970, who operated as Brisco Sawmills Ltd. until the mill was destroyed by fire in 1983. George Lautrup operates a wood preserving plant on the millsite known as Brisco Wood Preservers Ltd.

Edgewater Sawmills Ltd.
Harry Moore

In 1942, Harry Moore converted his portable sawmill to a stationary mill and planing operation at Edgewater. The mill operated for 24 years and was then sold to Revelstoke Sawmills (Radium) Ltd. in 1966.

Tom Alton & Sons Ltd.

Tom Alton & Sons Ltd. started a sawmill in 1928 at Parsons. It operated for 22 years, selling out to Cranbrook Sawmills in 1950.

Golden Lumber Co. Ltd.
(Also Kicking Horse Forest Products)
Henry Sigalet and son Jack

In the late 1940s, pole-maker Henry Sigalet started Golden Poles Ltd. in the Golden area. He later changed the name to Golden Lumber Co. Ltd. and in 1957 hired his son, Jack, to manage the operation. In 1960 the Sigalets sold their mill to an American firm. Renamed Kicking Horse Forest Products Ltd., the operation expanded into veneer and plywood production. In 1969 it was sold to Evans Products Ltd.

Evans Forest Products Limited
Unusual beginnings

Evans Products Company's unusual entry into the lumber business began with bicycle manufacturing and car battery production. The company needed yellow cedar to separate the batteries' lead plates. A sawmill at Coos Bay, Oregon, was the start of Evans lumber operations.

In 1928 Evans bought a mill on the north arm of the Fraser River in Vancouver, at the south end of Main Street. Lumber was produced at this mill, along with stock for battery separators which were produced at a nearby remanufacturing plant. After the sawmill was destroyed in a 1948 fire, Evans built a plywood plant on the site in 1954 and consolidated the reman plant at the same location. The plywood plant was sold in 1972, after 18 years of operation.

Meanwhile, Evans had expanded into the interior of B.C. In 1968 the Company purchased SAVONA TIMBER CO. LTD., on the west end of Kamloops Lake. It later also acquired COMMERCIAL LUMBER COMPANY in Lillooet, Kicking Horse in Golden and SELKIRK SPRUCE MILLS in Donald. In 1986, following an economic recession period, the parent company in the United States was forced into bankruptcy and the B.C. operations were put up for sale.

Except for its Thompson Valley and Cariboo holdings, Evans' Canadian operations were bought by nine senior managers. Under the new name, EVANS FOREST PRODUCTS LTD., the company concentrated its operations in the East Kootenay with a sawmill at Donald and a plywood plant in Golden. In 1993 Evans purchased WESTAR TIMBER'S Eagle River operations.

A former company executive, FRED MILLS, was President of the ILMA in 1984-85. Company President ED DORION was ILMA Chairman from 1991-93 and remains active on the ILMA Board of Directors.

Selkirk Spruce Mills Ltd.

During the Depression decade, pioneer sawmiller JIM PARKIN earned his living by sawmilling in Pine Creek, near Donald. In 1940 Parkin moved his mill to Donald. Three years later he sold it to Brick Chapman who named the company SELKIRK SPRUCE MILLS Ltd. Chapman sold to WILLIAM PIDGE in the early 1950s. Pidge operated until he went bankrupt in 1960. The same year the bankrupt mill was sold by a receiver to a consortium of owners, FRED BROEDER, HAP SIMPSON, PINK MILLER and silent partner, JIM RASKIN, a resident of California. DICK GONDEK joined the company as a full partner in 1961 and managed the operation until its purchase by Evans in 1970.

"Lumberton, B.C." A community that once existed southwest of Cranbrook.

Lumberton, 1927-1941.

BC Spruce Mills, Ltd., Lumberton, B.C. 1924-1941

BC Spruce Mills Ltd., Lumberton, B.C. 1924-1929.

BC Spruce Mills, Lumberton, B.C.

Flume tunnel, 1924-1940.

BC Spruce Mills, Lumberton, B.C. 1924-1941.

"Recollections"

"I remember when we used to work without hard hats. We didn't even own a hard hat. No hard-toed shoes; no rollover canopies (on machinery). Safety regulations have saved a lot of lives."

– Ross Gorman, (Managing Director, Gorman Bros. Lumber Ltd.)

"I was driving the 'Cat, making roads in the spring of the year and the sidehill was pretty steep at that place. The 'Cat rolled over twice and I scrambled out and never got hurt. I was lucky that time."

– Ken Long, (Sole surviving ILMA founding member)

"In the 1950s I was on the ILMA Board of Directors and a member of the forestry committee. We had a meeting with the Premier (W.A.C. Bennett) and other members of the Cabinet. We wanted to address the question of stumpage rates on cedar which we felt were too high. We all got dressed up in shirts and ties for this meeting with the cabinet. Before we got started the Premier lowered his glasses and looked up and said, "Gentlemen, this is the most prosperous looking group that we have interviewed today." And that was before we even had a chance to read the brief. Mr. Bennett just turned to Wesley Black (a member of the Cabinet) and said, "What is the next thing on the agenda?" We might as well have saved our breath."

– Warren Larson (Retired General Manager, Bell Pole Co. Ltd.)

Original Kelowna sawmill crew.

Gold Creek, 1944.

The ILMA's First 52 Years

Introduction

After several exploratory sessions, southern interior lumber producers met in Vernon for the inaugural meeting of the Interior Lumber Manufacturers' Association (ILMA). The founding date, August 21, 1941, was shortly after the start of the Second World War.

Forest companies had at last created a vehicle to carry a unified voice for logging and sawmilling concerns in the southern interior of B.C.

There had been an earlier attempt to maintain an umbrella organization for the interior lumber industry. A forerunner of the ILMA, known as the MOUNTAIN LUMBERMEN ASSOCIATION, was put together before the First World War by a coalition of producers from Kamloops, the Kootenay region and Calgary, Alberta. The headquarters, originally in Nelson, was later moved to Calgary. The poor market economy brought on by the Great Depression cut short this organization's history.

Common Concerns

With the disbanding of the Mountain Lumbermen Association, a vacuum had been created, and the motivation to form another industry association began to build again. Small sawmills operating individually knew they could not have much input into such important business-related matters as stumpage and freight rates, forestry rules and licensing regulations. Reviving an industry association appeared possible again as the war economy brought the Depression to a close and rejuvenated the market for lumber.

Impetus to form an interior industry association also came from an early federal agency known as TIMBER CONTROL.

Eight companies, all located in the Thompson/ Okanagan, formed the first ILMA membership. Ten years later, before the era of consolidation, membership had grown to 130 sawmilling and 30 associated companies. Since then, membership has included mills in the Thompson Valleys, the Okanagan and Similkameen Valleys, the Shuswap and Revelstoke regions, the Kettle Valley and the East and West Kootenays.

With the exception of the PATCHETT LUMBER COMPANY in Quesnel, all ILMA member companies operated in the forests and communities of the southern interior. Patchett remained with the ILMA even for a few years after the establishment of a Cariboo regional association, the CARIBOO LUMBER MANUFACTURERS' ASSOCIATION (CLMA).

Throughout the 52-year history of the ILMA, issues of most pressing concern to its membership have changed continually. To best represent the interests of member companies, the ILMA's agenda has stayed current with predominant issues of the times. Alternately or consecutively, the Association has dealt with rail freight rates and occupational health and safety matters, stumpage and royalty rates (paid to government for timber cutting rights), harvesting practices, reforestation and environmental requirements, lumber grading, log scaling, community relations – an endless stream of forest-related issues.

Mountain Lumbermen Association Annual Meeting, Nelson, B.C. 1911.

Presentation of National Safety Council Award to ILMA, 1965. (Accident rates have been drastically reduced since the forest industry's early days.) Left to right: "Scotty" Allison, Director of National Safety Council; Ian McDiarmid, ILMA President; Mert Bell, Chairman of ILMA Accident Control Committee; Al Macdonald, ILMA Manager; Ed Morrissey, Senior Safety Director for ILMA. (Photo by Paul Ponich Studios, Kelowna, B.C.)

INTERIOR LUMBER ILMA MANUFACTURERS' ASS'N

ILMA Founding Members, 1941.

Billy Smith,
Armstrong Sawmills

Ken Long,
Long Bros. Sawmills

Harry Turner,
Kamloops Lumber Co. Ltd.

Wilfred Hanbury,
Pondosa Pine Lumber Co. Ltd.

Henry Sigalet,
H. Sigalet & Co. Ltd.

Ormie Harris,
Shuswap Lumber Co. Ltd.

Stan Simpson,
S.M. Simpson Limited

Jim Strother,
Vernon Box & Pine Lumber Co. Ltd.

With so many men in uniform, sawmilling companies had to advertise in order to get workers for local industry.

World War Two created an unprecedented demand for dimension lumber. Many interior sawmills suddenly had huge lumber orders to fill. But they encountered a new problem that made it difficult to fill the orders. An acute shortage of labour was created as most of the country's able-bodied men were in uniform and committed directly to the war effort. The labour shortage was alleviated partly by another war-related event, the re-location to the interior of thousands of Japanese-Canadians from the Pacific coast. Many of these industrious workers were employed in the wartime and post-war forest industry.

Among commodities rationed in Canada during the war were gasoline and motor vehicle tires. Fred Stinson, while researching the industry's past, found that forestry vehicles accounted for a mere 3 percent of vehicles in the province during the war years. Yet the forest sector was supplied with 45 percent of the tires sold in B.C. Lumber was obviously of prime importance to a country at war. The thousands of men going into the service had to be accommodated in wooden barracks and camps that had to be constructed in a great hurry.

In April of 1943, the forest industry asked the Federal government for help in two areas. The first was a request to exempt sawmill workers from military service. The second, also of prime importance, was a protest that the meat ration, two pounds per man per week, was inadequate for loggers and sawmillers. They presented the case that the meat consumption in logging camps in British Columbia was a necessary eight pounds per man per week. As far as can be determined, neither request was approved.

Lumber Tariffs – Ongoing Concerns

> *"It has been a bitter battle trying to convince our American friends that we are not subsidized. It has been tested in the courts a number of times and they have been overruled on it. They keep coming back with it again. It has been a very frustrating part of the business in the last 10 years."*
>
> – ABE FRIESEN, (PRESIDENT, POPE & TALBOT LTD.)

Lumber tariffs, still looming over Canada today, have cast a shadow over the forestry industry since the earliest days of the ILMA.

American lumber producers have long resented the import of B.C. lumber into a marketplace they have preferred to consider their territory. Through the years American producers have maintained pressure on their government to have tariffs imposed on Canadian lumber. Their repeated efforts have usually been based on the argument that stumpage rates in B.C. amounted to a subsidy of B.C. operations. Their contention has led to allegations of unfair competition with American producers.

In December of 1961, American producers lobbied vigorously against Canadian lumber imports. They asserted the opinion that, "The government of B.C. is subsidizing its lumber manufacturers in the purchase of provincial timber in order to assure security of investment and expansion of employment."

Tariffs and/or surcharges have been imposed on several occasions in the past, resulting in added costs to the B.C. producers. Access to American markets has been obstructed as long as the tariffs have remained in place. A U.S. tariff, or countervailing duty, currently in place, is still being debated.

The Beginnings of Sustained Yield

Another significant historic event to emerge from the ILMA's early years was a recommendation put forward in 1943 by the ILMA, along with other forest associations and interested sectors. The common recommendation was that the government develop and adopt a sustained yield policy in the management of provincial forests.

The extraction and sale of any resource commodity has always been market driven. The early wood product market, glutted with resources of all kinds, assigned market value only to the finest, highest grades of lumber. Timber of inferior quality was burned or left in the bush, simply because no one wanted to buy it. It was no secret that market-driven forest practices created enormous "waste" of the resource. Many loggers and sawmillers were appalled by the wasteful practices.

Concerns about future timber supply prompted industry to pressure government for change. The pressure did not come from citizen groups, for during the 1930s and the war years, there were virtually none in existence. Concerns expressed by the ILMA, its counterpart associations and others, led to the setting up of a Royal Commission into forestry, led by GORDON SLOAN, then Chief Justice of the B.C. Court of Appeal. From the establishment of the Sloan Commission, the goal of sustained yield has gradually been given greater emphasis by a series of provincial governments.

HARRY TURNER, the first chairman of the ILMA, was instrumental in urging changes in provincial forest policy, particularly with regard to better conservation and utilization of the forest resource. Sloan invited Turner to act as an advisor to the Commission in matters pertaining to the interior lumber industry. Turner's valuable contribution to the Commission's work resulted in the embodiment of many of the recommendations of the ILMA.

Railroad Freight Rates

The long-running issue of rail freight rates dogged the ILMA and its member companies through several decades. Southern interior forest companies traditionally relied heavily on markets in the prairie provinces and eastern Canada. Before trucks moved in as competition, railroads were the sole means of transport to supply lumber to these markets.

In these pre-trucking years, minutes of ILMA annual meetings were rife with references to freight rate problems. In 1946, railroads were advocating a 30 percent rate increase. This increase would come at a time when lumber prices were frozen by wartime regulation. (Wartime price controls were not lifted until 1947.) In 1949, the ILMA membership calculated that rail freight rates to Alberta accounted for 20 percent of the total cost of lumber to the province's consumers.

As the rail freight issue dragged on, then ILMA President Harvey McDiarmid, a war veteran, said the ordeal reminded him of the Allied landing at Normandy ten years earlier. The odds against surviving the battle seemed roughly similar.

The construction of an improved highway system through B.C.'s challenging terrain, coupled with the evolution of high-capacity lumber trucks, made rail service less critical. Lumber companies no longer had to rely totally on one means of transportation. Since the construction of the Rogers Pass section of the Trans-Canada Highway, a continuous stream of lumber trucks has passed through, bound for eastern markets.

Canada's de-regulation of railroads in recent years has allowed individual companies to negotiate more competitive shipping arrangements. Although the abandonment of railroad branch lines also created more opportunities for trucking, rail is still the mode of transportation for the largest percentage of lumber exports. Southern Interior lumber continues to move smoothly to markets by both rail and truck.

Lumber Grading

Another transformation was in store for the interior forest industry as it progressed from pioneer days to the sellers' markets of the Second World War, and finally, into the highly competitive post-war marketplace.

From the industry's early days, lumber products were graded, priced and sold according to quality, appearance and structural characteristics. Early mills produced lumber primarily for local customers, and it was acceptable to establish grades of lumber informally, between the mills and the customers. This community-level approach was custom-fit to the circumstances of the time.

As mills increased in size, and as transportation systems evolved, mills were able to reach markets at greater distances. It became increasingly apparent that common lumber standards needed to be established between mills and customers in expanding trading circles. The need for a standardized approach to lumber grading had been one of the key forces behind the formation of the ILMA.

In October of 1946, interior B.C. companies were elated to find a market for their spruce in South Africa. But their celebration was short-lived. South African buyers advised that Finland and Sweden were providing better-made, better-graded lumber stock. Access to overseas markets was becoming more competitive, and quality was the competitive edge. The South African experience underscored the need for the ILMA to establish quality control and grading standards, and for member companies to produce uniform, quality products.

In 1951, the ILMA undertook dual initiatives aimed at standardizing lumber products. The first was to prepare a common training program for lumber graders. The second, to develop and institute an inspection program. ILMA quality control staff assumed the responsibility of inspecting the grading quality at member mills. The diligent application of standards ensured that grading decisions were made within established tolerances.

The following year, the Association's first lumber grading courses attracted 264 trainees throughout the southern interior region. Just four years later, the number of graders attending classes had increased to 465, a turn-out which clearly illustrated the growing importance of quality control.

Another important development in the drive for standardized lumber grades began in 1956 with the concept of an ILMA grade stamping system. An "association specific" grade stamp became a reality the following year, when six member companies were certified to apply this grading mark to their lumber production.

United States' acceptance of the ILMA's grade stamp was given a major boost in 1960. The U.S. Federal Housing Administration officially recognized the Canadian Lumber Standards Accreditation Board. This Board had, in turn, recognized the grading standards of the ILMA and a number of other Canadian lumber grading agencies. This success was followed up, a year later, with official recognition of the ILMA grade stamp by the Southern Building Code Congress, representing 13 southern U.S. states. These were undoubtedly the most significant developments to that point, in terms of standardized lumber grading in the North American marketplace.

Beginning in the early 1960s, however, pressure was growing in the U.S. to replace the various lumber grading standards with a single standard, that of the U.S. National Lumber Grading Committee. In response, the Canadian industry established the National Lumber Grading Authority (NLGA) to develop a single standard for Canada, known as the NLGA Rule.

A milestone was marked in 1970 when the current ILMA President, Bob Thom, announced at the annual meeting that American Lumber Standards (ALS) had officially approved the new Canadian standard for use in the U.S.

The ILMA maintains an active role in keeping lumber grade requirements current with the needs of rapidly changing international markets. The interior industry's consistently high standard has expanded its market opportunities, most notably in the U.S., Europe, Japan and Australia.

Presentation of Senior Grading Trophy, 1965
Pictured from left, Bob Thom, Shorty Omatsu and Cliff Hearn

25 year reunion of the founders of the ILMA. Left to Right: Ormie Harris, Jim Strother, Ken Long, Horace Simpson (son of founding member Stanley Simpson), Jack Smith (son of founding member Billy Smith).

1972: The Record Year

In 1972 the biggest building boom in North American history reached its peak. Lumber prices also reached a new peak with the coming of age of the large number of World War Two baby boomers.

Building activity increased dramatically in Canada and in the States. The U.S. reported 2.3 million housing starts and the construction of half a million mobile homes in that record-breaking year. Canadian industry shipped 9 billion board feet of lumber to the U.S. market, one quarter of all the lumber used in the United States in 1972. ILMA mills accounted for 1.4 billion board feet of that production.

Within the decade that followed this extraordinary growth trend, U.S. protectionist sentiment was once again directed toward Canadian lumber producers.

"The most exciting time to be in the industry was in the mid-to-late 70s. Expansion followed expansion, markets were steady and the industry was going great guns. I never expect to see those days again. Since the 1981 recession the field of industrial relations has been a completely different world. The recession changed attitudes and the industry never bounced back to the expansion and growth mode. The industry chopped and trimmed and has been running lean ever since. Technological changes had to happen at all levels to keep the companies competitive in the marketplace. Maybe it was too easy before and the industry will be tougher and stronger in the end."

– Wayne Nowlin (President, IWA Local 1-405)

The Pearse Commission

Headed by DR. PETER PEARSE, B.C.'s third Royal Commission into forestry began in 1976. The ILMA presented a brief to the Commission, highlighting the problems created by increasing alienation of forest land to non-forestry purposes.

Implementation of some of the Pearse recommendations, in the late 1970s and early 80s, gave B.C. a new Forest Act. This was followed up with a total revamping and reorganization of the Forest Service. Equally dramatic changes in forest regulations demanded increased Association time and effort on matters such as environmental protection, silvicultural responsibility, land use, and many other issues. Federal legislation, relating particularly to the protection of fish habitat, was also strengthened during this tumultuous period. The business of forestry would never again be simple.

"If it were not for computers we would have to have a ten-story office building."

– GORDON STEELE, (PRESIDENT & CEO, RIVERSIDE FOREST PRODUCTS LTD.)

From Imperial Measure to Metric

Metric conversion was drawing closer. In 1967, Art Lloyd, chairman of the Canadian Wood Council, accurately forecast the imminent need to cut lumber to metric dimensions that were accepted in most international markets.

As the drive to conversion moved ever closer, the ILMA struck a committee to specifically address metric conversion and its ramifications. Jim Charpentier, Chairman of the Metric Committee, reported at the 1976 ILMA annual meeting that the target date for beginning shipment of metric lumber was September, 1979. The Canadian forestry sector's target date for Canada-wide completion of metric conversion was July 1, 1979.

The conversion to metric measure was a major cost and dreadful inconvenience for wood producing companies, especially smaller ones. On the long list of complaints was the fact that complete, and likely favored, sets of tools would suddenly no longer fit the metric-sized nuts and bolts.

North American builders were accustomed to buying lumber in Imperial dimensions and many markets wanted to continue having it supplied in feet and inches. The U.S. preference for Imperial-sized lumber convinced many interior operations to cut lumber to both dimensions.

Recession and Restraint

In 1980, a major concern of ILMA President GORDON FISHER was the lumber industry's struggle to cope with rising inflation and steadily increasing costs. The early 80s economic recession had caused a sudden and dramatic decline in housing starts. The slowdown in construction directly affected ILMA markets, both in Canada and in the U.S.

In his second year as President, Fisher was still not encouraged by the economic climate. The downcycle had continued unabated, mainly because inflation and high interest rates had boosted housing costs beyond the reach of many prospective buyers. Lumber markets, and many mills along with their employees, suffered accordingly.

Predictably, as the U.S. and Canadian lumber demand declined, a coalition of American lumber producers renewed the protest against the import of lumber from Canada. American producers stepped up pressure on their government to impose a COUNTERVAILING DUTY on Canadian lumber. Coming at a time of severe recession, an American duty posed an ominous prospect for many Canadian lumber producers.

The Chairman of the ILMA's Logging and Forestry Steering Committee predicted that the implementation of the provincial government's restraint programs in 1983 would directly impact the interior industry. The impacts related to the down-sizing of Forest Service staff and the transfer of greater levels of forest management responsibility from government to the private sector.

In an unsettling economic climate, in 1986 B.C. had a succession of three provincial forests ministers. The same year, a 15 percent export duty was imposed on most Canadian lumber being shipped to the U.S. The following year, the export duty was replaced in B.C. with a corresponding increase in government stumpage rates.

Other significant changes in government policy occurred at the same time. The Forest Service sold its nurseries to the private sector. It assigned the financial responsibility for basic silviculture, including tree planting, to the holders of Tree Farm and Forest Licences. And it removed 5 percent of the allowable cut from these licences to establish a SMALL BUSINESS PROGRAM.

As the 1980s came to a close, the B.C. Southern Interior forest industry had grown highly dependent on export markets. In 1989, 72 percent of ILMA lumber shipments were exported – 67 percent to the United States and 5 percent overseas, with the remaining 28 percent to domestic destinations.

The Great Environmental Debate

> *"Another thing that is good is public awareness. I know, as foresters in the past, we could not convince anybody we had to spend more money on forestry and silviculture. I think any politicians today who said they were going to cut down on tree planting would be sounding their own death knell. Now the public will insist that it happen. I guess it is thanks to the environmental movement that it is now a high priority."*
>
> – ED DORION, (PRESIDENT & CEO, EVANS FOREST PRODUCTS LIMITED)

> *"We had to build 40 miles of road to open up the country to the north and east of the (Okanagan Falls) millsite. That was in the early 1970s. I remember saying to some of the more vociferous environmental protesters that if they would agree to come back in 20 years they would see a fine new forest thriving there. I have been up and looked at it and I feel really good about it. The new trees are 20 or 30 feet tall."*
>
> – FRED STINSON, (RETIRED ILMA PRESIDENT, 1977-79)

> *"Twenty or 25 years ago the environmental disputes started. They have been going steady since then. Originally, we employed two people in the forestry department and now we need five, and that doesn't take into account the other people going to meetings etc. That is just in the logging department. A lot of what we do is now better, but a lot of it is environmental overkill."*
>
> – JACK WIGEN, (PRESIDENT, WYNNDEL BOX & LUMBER LTD.)

> *"I remember the debate over logging in the Blewett watershed south of Nelson. There were about 150 domestic water licences up there. In 1975 we had a public meeting and 300 people voted 'no' to any logging in the watershed which was the headline in the Nelson Daily News the following day. So we formed a watershed committee and after giving people some guarantees about water quality and water quantity, we turned it around and were given full approval to cut the resource. And this happened in Nelson, the environmental hot-bed of the Kootenays. One of the things I remember about it was, as soon as it looked like we were going to work things out, the media lost interest in us. At least until about four years later, when the Nelson Daily News wrote a headline saying the Blewett watershed was a model of co-operation in watershed management."*
>
> – JOHN MURRAY
>
> (RETIRED V.P., WOODLANDS DIV., C.F.I.) DESCRIBING ONE OF THE INTERIOR'S FIRST PUBLIC INVOLVEMENT PROCESSES

As the ILMA neared its 30th anniversary, B.C. began to feel the presence of other groups with an interest in the forests. Non-commercial groups and interests began to exert influence on forest management and land use policies.

Traditionally, concerns about the ability to maintain forest values such as wildlife populations had related mainly to game species. Wildlife objectives could usually be accommodated by commercial forestry within the framework of multiple or integrated use. In some cases, big game numbers, especially elk, moose and deer, actually increased after logging. But as society entered the environmental era, wildlife concerns broadened to include the full range of species, some with habitat requirements that could be accommodated only in mature or old growth forest environments. Through the 70s and 80s, environmental objectives broadened even further to include withdrawals of large tracts of land for such values as viewscapes or wilderness.

> *"Over the years, environmental interest has gradually increased. Especially over the last four or five years. It has made our company do a lot more planning and a lot more careful logging. We have gradually adapted to it. It has been good for the industry in*

some ways, and in others it has decreased the areas we can operate in. I guess that is just the cost of doing business today."

– Bud Nelson, (President & General Manager, Galloway Lumber Co. Ltd.)

"There is no one who likes fresh air and clean water better than I do. It seems the squeaky wheel gets the grease. I guess environmental issues have been going on for quite a while. It is good to have a certain amount of that going on, but there is also a lot of B.S. now and it has gone so far they are not using good judgement. We have to look after our forests, it is part of our economy and it is what keeps this country going."

– Ken Huscroft, (President, J.H. Huscroft Ltd.)

Throughout the history of the industry, a secure and adequate wood supply has been a universal need of forest companies. The ability to obtain financing has often been tied directly to the security of wood supply. The possibility of logging being restricted or downright prohibited on an increasing number of sites created well-founded anxiety throughout the industry.

Campaigns for non-use, or non-commercial use, of forest land have now become commonplace. The ILMA, along with guides, ranchers, miners, and other commercial user groups in the forest, continues to cope with an accumulation of changing demands. The ongoing process of resource management planning continues to create much uncertainty about the forest industry's future.

"When I first came to the Rocky Mountain Trench in 1954 there were only two major parks. The working forest land was all the rest of it. We could move almost anywhere without restriction. Since that time I have seen a lot of the land base change, not only because of ecological reserves, parks and wilderness, but also because towns and roads and railroads and dams, real estate and recreation developed. All this has taken place in the last 40 years. We have effectively been downsized. Our freedom to move on the land is being eroded and taken away a bit at a time."

– Carmen Purdy, (Manager, Community Relations, C.F.I.)

"The Shuswap area attracts a lot of Alberta tourists who are here for a week or two, or perhaps they may own a cabin on the lake. These cabin owners are absentee residents. Often they don't want any logging or industrial activity within their recreational sphere. Federated Co-operatives Limited does a lot in attempting to appease them. Meetings have been held in Calgary with cabin owners, to explain five-year development plans for the area. Often when our logging plans are explained to them, the common response is "Is this all you are going to do?" Their reaction confirms that someone has intentionally fed them "factoids" or misleading information."

– Charlie McCarthy, (Senior Vice President, Federated Co-operatives Limited)

"There is no question that on a world-wide basis the timber resource is a diminishing resource and we need to be better stewards of it. I do think that some of the things we do tend to swing to the extreme, but we now have a trend to better utilization of the resource. I like that trend. As painful as it sometimes is, I think it is a good trend."

– Abe Friesen, (President, Pope & Talbot Ltd.)

Entering the 1990s

Uncertainty and unpredictability marked the interior forest industry's entry into the final decade of the 20th century. As the decade opened, inflation had been replaced by a period of lower interest rates, but declining prices. President John Marritt reported to the 1990 ILMA convention that lumber prices had dropped to 1981 levels, and that chip prices had fallen with a decline in the price of pulp.

The interior industry faced an ever-increasing series of demands on the forest land base. Lists of conditions and requirements raised operating costs to a point that came close to matching, and in some cases exceeding, shutdown costs. In the last quarter of 1990 these various burdens resulted in curtailed sawmill shifts, implementation of work-sharing programs, extended Christmas holiday closures and in some cases, total shutdown of mills. A 12 percent decrease in lumber shipments, compared to those of 1987, was reported. A few years into the 90s, markets for lumber had grown stronger.

The 90s also saw several government initiatives established to address short and long-term problems of environmental concern, land use and resource allocation. Such bodies as the FOREST RESOURCES COMMISSION, THE B.C. ROUND TABLE ON THE ENVIRONMENT AND THE ECONOMY and THE COMMISSION ON RESOURCES AND ENVIRONMENT (CORE) were set up with varying, and sometimes overlapping responsibilities.

The ILMA and its member companies, along with the full range of citizen and industry organizations, found its time and resources stretched to the limit in order to take active roles in these various initiatives dealing with present and future forest land use in B.C.

THE B.C. ROUND TABLE ON THE ENVIRONMENT AND THE ECONOMY, a body of mostly volunteer citizens from environment, labour, industry and Native Indian groups, was given the mandate to advise government on directions for "sustainable development" of the environment, the economy and communities in B.C. This body, a local offshoot of the United Nations Commission on Environment and Development and the National Round Table of Canada, was responsible for adding a new dimension to the land use debate in B.C. The term "sustainability" has now become part of the popular language in strategic and operational land use discussions throughout the province.

After a series of town-to-town hearings throughout the province, the Round Table produced a document verifying that forces concerned with business, the environment and community sustainability were on a collision course. To avoid increasing conflict, the Round Table advised a consensus-based plan for overall sustainability was essential.

The provincial government also moved to deal with valley-by-valley land use conflict by creation of the COMMISSION ON RESOURCES AND ENVIRONMENT (CORE), headed by former provincial Ombudsman STEPHEN OWEN. The CORE process is currently wrestling with the many serious land-use conflicts in hot spots – Vancouver Island, the Chilcotin-Cariboo and the Kootenays. The government, industry and the public are counting on these ongoing processes to develop a workable provincial LAND USE STRATEGY.

THE FOREST RESOURCES COMMISSION, which operated in the early 1990s, and the B.C. Round Table produced studies on the level of dependence of various communities on forest industry activity. For the first time it was possible to quantify just how important forest activity is to the social and economic well-being of dozens of interior towns. In many cases, the economies of towns within the ILMA's area of interest were revealed to be very highly dependent on forest industry activities. Such communities as Cranbrook, Golden, Falkland, Merritt, Princeton, Grand Forks and many others, emerged as still relying on forestry for the largest percentage of economic activity and employment.

The Falldown Effect and the "netting down" of the harvest

As the industry entered the 1990s, the so-called "FALLDOWN EFFECT" began to manifest itself. The falldown effect, a natural phenomenon in any forest situation, occurs when old high volume original stands are replaced with second growth. Second growth forests usually have less volume than old growth because they will be harvested at an earlier age. Obviously, this falldown in inventory will require a corresponding drop in the allowable cut.

The inevitability of the falldown effect had been emphasized in the Royal Commission report of Peter Pearse in 1976, but there had been discussion about it long before. When falldown was first talked about, the great debate over assigning mature forest to non-timber purposes had not yet begun. Compounding the cut reduction due to falldown, is the "NETTING DOWN" of the cut to accommodate more non-timber uses and the increasing loss of forest lands to parks, reserves and preserves. In the 1990s and for the foreseeable future, timber supply will remain controversial as cut reductions due to natural falldown are compounded by "netting down" the forest for more non-timber purposes.

Needless to say, large reductions in the cut will create significant economic and social impacts for the companies, their employees and the communities involved. Since forestry also comprises one of government's largest sources of revenue, there is bound to be a negative impact on government spending as well.

Thus, in a welter of changing government regulation and public demands, the ILMA completed its 52nd year. The effect of dynamic social, environmental and economic demands on the province's land base creates considerable uncertainty about future timber supply.

"Through the 28 years I served with the ILMA, the competent, very able staff made my job much easier and their assistance was much appreciated."

AL MACDONALD
(AL JOINED THE ILMA IN 1956 AS ASSISTANT MANAGER, AND IN 1958 BECAME MANAGER, SERVING IN THAT CAPACITY UNTIL 1984.)

ILMA members – 1950. (Photo by Industrial Photographers, Vancouver, B.C.)
Standing: Left to Right: W.J. Smith (Armstrong Sawmill Ltd.); H.B. Simpson (S.M. Simpson Ltd.); G.W. Eriksson (Knight Lumber Co.); Jas. White (Wilson Lumber Co.); George Perry (Summerland Box Co.); Harry Burns (Burns Lumber and Coal Co.); O.S. Harris (Harris Lumber Co.); Walter Jostad (Cranbrook Cartage and Transfer Co. Ltd.); V.C. Brown (Cranbrook Sash and Door Co. Ltd); J.B. Miller (Thompson Valley Lumber Co. Ltd.); Boris Turin, (Zamora Forest Products Ltd.); F.W. Kickbush (Princeton Box Co.); W.G. Spence (Cranbrook Sash and Door Co. Ltd.); V.G. Smith (B.C. Division, Canadian Manufacturers' Association).
Seated: Left to Right.
K.A.W. Long (Kamloops Lumber Co. Ltd.); N. Beech (Portable Sawmills Ltd.); A. Perrault (Rutland Sawmills); J.W. Munsie (Kettle Valley Lumber Co. Ltd.); J.G. Strother (Vernon Box and Pine Lumber Co. Ltd.); W. Hanbury (Pondosa Pine Lumber Co. Ltd.); H. McDiarmid (Timber Sales and Distributors Ltd.); S.M. Simpson, Chairman of the ILMA Board; (S.M. Simpson Ltd.); R.A. Mahoney, Secretary, ILMA.; Hugh Dalton (Manager, B.C. Division, Canadian Manufacturers Association).

ILMA Chairmen/Presidents 1941 to Present
(The elected heads of the ILMA Board)

Harry Turner, Chairman 1941-1944

Ormie Harris Chairman 1944-1946

Gordon McMynn Chairman 1946-1948

James White Chairman 1948-1949

Stanley Simpsom Chairman 1949-1951

Vic Brown Chairman 1951-1953

Harvey McDiarmid President, 1953-1955

Gus Eriksson President, 1955-1957

Vern Vance President, 1957-1959

Art Lloyd
President, 1959-1961

Jack Munsie
President, 1961-1963

Ian McDiarmid
President, 1963-1965

Dennis Brookes
President, 1966-1967

Jim Charpentier
President, 1967-1969

Bob Thom
President, 1969-1971

Henry Nelson
President, 1971-1973

Ed Drew
President, 1973-1975

Marvin Persson,
President, 1975-1977

Fred Stinson
President, 1977-1979

Gordon Fisher
President, 1979-1981

Gerald Raboch
President, 1981-1983

Fred Mills
President, 1983-1985

Fred McLellan
President, 1985-1987

Mike Rouse
President, 1987-1989

Ross Gorman
Chairman, 1989-1991

Ed Dorion
Chairman, 1991-1993

Morris Douglas,
Chairman, 1993-1995

ILMA Annual General Meeting – 1964. Prince Charles Hotel, Penticton
Left to right: Mrs. J.W. Munsie (Dorothy), Mrs. A.P. Lloyd (Gloria), Mrs. A.D. Macdonald (Bernice).

Some Recollections About Women and the ILMA

by Bernice Macdonald

Bernice Macdonald is well known for organizing many fine social events over the years while her husband (Al) was with the ILMA (1956 to 1984).

My introduction to the Ladies' program was at the Annual General Meeting in April of 1957. On Wednesday evening the wives of the Directors met for dinner in the Prince Charles Hotel in Penticton while the Directors had their Board meeting and dinner. At that time most of the sawmills were independently owned. Most of the Directors and their wives were from small towns scattered throughout the southern interior of British Columbia and they were a very congenial group.

Marge Rees, wife of Secretary-Manager, Jim Rees, was hostess for the occasion. There were probably about 10 ladies at the dinner. I remember meeting May Kirkland, whose husband, Fred, was the co-owner of Lumby Timber. I remember Evelyn Jostad too. Her husband, Walter, headed up Cranbrook Cartage and Transfer.

On the Thursday evening there was a cocktail party and smorgasbord hosted by Simond's Saws Ltd., represented by Don Weston, Jim and Mary Wallace, and Phil and Betty Quelch. Live lobsters were flown in from New Brunswick, a great treat for the western lumbermen and their ladies! This evening was always a pleasurable one. In later years the buffet dinner was followed by a no-host dance at the Penticton Golf & Country Club – good fellowship and good fun!

On Friday, twelve ladies met for tea at Clara's Tea Garden. Hats and gloves were the order of the day. That evening, there was a cocktail party, banquet, presentation of safety awards, and dance. A sing-song ended the evening with MARGE REES at the keyboard.

The official approval of the ILMA grade stamp in 1960 resulted in a marked increase in membership and also in attendance at the Annual General Meetings. To encourage greater participation by the ladies, I instituted the practice of sending letters detailing the ladies' program directly to home addresses rather than to the sawmill offices. We had fashion shows, fur shows, craft shows, bowling, bus tours, mini-golf tournaments, bridges, teas and luncheons. The ladies responded enthusiastically. I believe attendance reached 125 at one of our fashion shows, a far cry from the original twelve in 1957!

ILMA Annual General Meeting – 1972. Saturday afternoon get-together at Al and Bernice Macdonald's home. At the piano: Aline Charpentier, Left to right: Shirley Thom, Pat Drew, Edgar Drew, Alice Brookes, Bob Thom, Teddy Vance.

WINIFRED MATHER'S "Accessory Story" was a particular favorite, as was her fur and hat show. One hat show was memorable because JOAN SIMPSON and ELISE CROOKS purchased a hat (I think for $35) between them and took turns wearing it on special occasions. During the mid-60s, I remember the ladies wore evening gowns and white gloves and we had receiving lines preceding the banquet. This elegant touch to the evening was instigated by Ian and DOREEN McDiarmid. I was sorry to see it go. For many years, there was a traditional toast to the ladies by one of the men. Usually the President's wife responded to this toast. This particular part of the program was eventually eliminated, ostensibly because there were too many shy ladies who could not be persuaded to reply.

At the Prince Charles Hotel, some of the ladies complained that there was too much 'sawdust' piling up on the dance floor. But they knew that the Annual General Meeting afforded the opportunity for sawmillers to discuss their problems. They accepted the fact that their husbands were going to "talk lumber" no matter what!

DOROTHY MUNSIE, who has attended ILMA General Meetings longer and more regularly than any other lady, knew from first-hand experience about "sawdust on the floor" because husband Jack was an enthusiastic participant in "off-dance-floor" discussions.

At the 1980 meeting, Gordon and EDYTHE HILL'S white poodle, Barnabas, strayed from their room in the Pilgrim House Motel. EDNA FISHER, wife of the current President, Gordon, made a valiant attempt, with my inept help, to capture Barnabas. In the process she was bitten on the finger, and was rushed to the Emergency ward at the Penticton Regional Hospital where the finger was stitched. This is just another example of ILMA ladies willing to perform above and beyond the call of duty.

On the other hand, we have GLORIA LLOYD and ALICE BROOKES, who were confronted by a bear on the golf course at Fairmont Hot Springs. They quickly ran to hide behind Art Lloyd, with Art loudly exhorting them to find their own route to safety and not to lead Bruin to him. Art is a good friend of mine, I know him well. It is not in character for him not to defend and protect the ladies. Therefore, I have to explain that on the hole before the one with the bear, Art's ball landed in the fork of a tree. The ladies insisted that he "play it as it lies". Art hadn't quite forgiven them. It wasn't an easy climb to that fork in the tree!

Another friend I remember with deep affection is ALINE CHARPENTIER. An accomplished musician, she played the piano for many sing-songs, not only at our ILMA gatherings, but at the Truck Loggers' convention, and at Canadian Wood Council meetings too. I think our best times were at home on a Saturday after the Annual General Meeting, when Alan and I entertained the Directors, guests and their wives. With Edgar Drew's beautiful tenor and Jim Charpentier's true baritone, Aline had us all singing harmony as if we had years of training. I know that she enjoyed it too. Her music made the day for all of us.

On one memorable Saturday, we had delegates from the Truck Loggers' Association, as well as others, for a get-together. A neighbor was babysitting a pet racoon (a large pet racoon). It escaped, and on hearing the music decided that it would join the party. I made the mistake of letting it into the house. It was very lively! It stole cigarettes and ate them right out of the package, overturned plants, charged around the house like a tornado and generally created quite a diversion. Staunch forester Fred Stinson, with the aid of other lumbermen, put him in the doghouse and nailed boards securely across the opening. The truck logger delegates to this day believe that the racoon was an "import" designed to add life to the party. Actually, we had a card from "Tiki" the racoon, a few days later, to thank us and to say how much he had enjoyed the visit to our home.

There have been many changes over the years, but also many happy memories – and many good friends!

BC Spruce Mills Ltd., Lumberton, B.C. 1924-1927.

BC Spruce Mills Ltd., Lumberton, B.C. 1924-1941.

The Loggers

Loggers in the pioneer days of the interior forest industry were, by necessity, versatile and pragmatic workers.

Early portable sawmills were family operations or small partnerships. Each member of the crew was normally expected to be able to perform any task that needed to be done, in the bush or around the mill. Loggers gained first-hand experience as sawmillers, and sawmillers learned to fell trees.

The shared activities fostered a high level of self-reliance. The best logger/sawmillers were just as adept at falling and bucking trees as they were at operating and repairing milling equipment, loading and unloading lumber, and driving teams of horses.

At portable mill sites, crews often felled and skidded trees until they filled the mill yard. They then laid down their saws and axes, and turned their attention to cutting lumber, until the pile of logs was gone. The crews then started again, cutting enough trees to replenish the supply for sawing into lumber.

Portable sawmills were not set up in towns, close to markets, but in forests, close to the trees. With characteristic practicality, early sawmillers could see no sense in hauling whole logs out of the woods. Only parts of the log were useful for lumber. Logs were sawn in the forest, the sawdust and slabs were piled or burned, and only lumber was hauled to town.

Portable mills were unsophisticated machinery – simply beds made from squared timbers, fixed with a circular head saw, a carriage to push the log through the saw, and a form of power plant. Most were diesel-powered, but the more rustic were powered by gasoline engines, usually old truck or car motors. Even farm tractors were called into service to power little mills, with flat belts and power take-offs. Mechanical modifications and combinations were limited only by the creativity and ingenuity of operators. The objective, to get a mill up and running, fired inventive spirits.

Road access to the forests and high-volume trucks eventually revolutionized and simplified the logger's job description. Although early forest access was primitive, and trucks were sorry contraptions, far below today's standard of safety, comfort and efficiency (not

Steam donkey loading rail cars, East Kootenay.

even equipped with stereo), the road was now clearing for stationary sawmills in towns.

Most modern loggers are more specialized than their pioneer counterparts. But in many forest companies, the "jack-of-all-trades" heritage is alive and well. Partners, operators and co-workers, often self-taught, can be skillful welders and heavy-duty mechanics, as well as experienced road-builders and mill-wrights.

Specialized machinery needed operators with specialized skills. Efficient feller-bunchers, which cut and pile tree-length wood for pickup by grapple skidders, have taken much of the heavy hand-work out of logging. But this logging machinery has its limits. It operates best on fairly level ground and is not suitable for use in big timber, nor on steep sidehills or rough, broken ground. Hand falling with chain saws is still the harvesting choice in many areas. And loggers have to be as physically fit as ever, to scramble through the slash and snow to hook chokers for skidders, dozers, line yarders or helicopters.

Loggers sometimes contend that today's high-speed saws and yarding systems actually require them to work harder. Some say "keeping up to the machinery" has created considerably more stress for today's loggers. In the quieter and slower-paced days of horse or tractor skidding and crosscut saws, at least partners could talk to each other as they worked. Many of the skills associated with earlier times, such as chopping and hand-sawing, are now kept alive only as a logger sport or exhibition. It is rare to find an experienced axeman younger than 40.

Most logging for southern interior forest companies is now carried out by independent contractors. Contracting crews range in size from two to three men, to as many as 30 or 40, depending upon the range of the contract. A "stump to dump" contract, for instance, requires a crew to build roads, landings and skid trails, to cut trees, to load trucks and to haul logs to the mill, or log dump. A "falling contract" however, is concerned only with the cutting of trees. One or more separate contracts would then cover yarding, skidding and hauling responsibilities.

As the sawmilling industry changed after the Second World War, logging operations went through a corresponding transition. Logging became an even more complex activity. New regulations governed diameter limits and stump heights. New rules and regulations were required by agencies such as Workers' Compensation Board (WCB) and Ministry of Highways. From the provincial Ministry of Forests and Ministry of Environment, and the federal Department of Fisheries and Oceans and Environment Canada have come a maze of regulations regarding water quality, soils, fish and wildlife habitat, viewscapes and on... .

Loggers in the southern interior were quick to realize that they could not communicate individually with these various regulatory bodies on the plethora of regulatory requirements. In 1958, many of them formed their own association, Interior Logging Association (ILA).

The ILA membership has grown to approximately 500 logging companies, representing nearly 3000 loggers, contractors, contract silviculture workers and log haulers. From an office located in Kamloops, the ILA represents loggers from south of Williams Lake to the U.S. border, and from east of Chilliwack to the Alberta border. Other logging associations represent logging contractors in northern B.C. and at the coast.

At one time, coastal and interior loggers were split along more than geographic lines. A long-standing rivalry existed between forest workers who logged in different tree stands. Legend and local pride, more than serious antagonism, created a perception that coastal loggers considered their work – with bigger trees – more skilled and daring than logging jobs in the interior.

Times have changed, and 35 years since its inception, the ILA has represented its members in matters as diverse as licensing and regulation of logging trucks, Workers' Compensation regulation, federal and provincial taxes on logging machinery, and standardization of

Advances in mechanization led to more specialized machines – Lumberton area.

log scaling methods. The ILA has also been actively involved in the Kootenay/Boundary CORE process, as well as in many other public "land use " processes.

Currently, the ILA maintains regular contact with counterpart organizations elsewhere in B.C. as well as in Alberta, Alaska, Washington, Idaho and Montana. The ILA also works with the ILMA on providing safety instruction for tree fallers. Despite advances in technology, falling timber is still a most dangerous business. Logging and sawmilling associations have actively contributed to resolving many work-related safety issues.

Today's interior loggers must master increasingly sophisticated equipment.The diverse nature of the working environment, ranging from near-desert to belly deep snow, calls for flexible and adaptable individuals.

Environmental concern over logging and harvesting methods has affected every region of the province. The politicizing of these concerns requires loggers, along with everyone in the industry, to learn a range of new skills. For many, these include public speaking, media relations and other communication skills. Virtually unnecessary just a few decades ago, communication with the public is destined to demand a greater share of loggers' attention.

Planned blockades of logging roads have occurred in a number of southern interior regions. In one sense, environmental opposition to certain aspects of logging has caused considerable social hardship for people who make their living in the woods. While society consumes wood products, logging as an occupation is denigrated in the media almost daily. Facing constant criticism of their life's work, traditionally a source of personal pride and self-esteem, has not been easy for many loggers and their families.

The fast-track of change has touched all aspects of forestry. Working in the woods near the end of the 20th Century is a far cry from simpler times – when the loggers' greatest challenges were to get a road in and to haul wood out.

Early days logging camp, Ross – Saskatoon Co., Waldo/Elko area.

Winter horse logging. Note the logs are "dogged" together on the skid trail. Locking the logs together in this way slows them down on steep grades and keeps them from running into the skidding horse or team. Yahk Lumber Co.

"When I was a teen-ager, I was working on a float camp on the coast (on steep coastal terrain, loggers were often housed in 'floating camps', on rafts tied to the shoreline). Once when I was asleep the others in the crew untied the raft, with my cabin on it, and turned it around so that the door was facing the water instead of the walkway. And they tied a rope about knee height right across the door. Then they hammered on the wall and shouted that I was late for work. I ran out the door and went head-first into the water."

– GORDON STEELE, (PRESIDENT & CEO, RIVERSIDE FOREST PRODUCTS LTD.)

"In the years when loggers lived in camps we were much closer and more involved with each other. We would go out to parties in some of these logging camps on a Saturday night. Great experience. What a great time we all had."

– HORACE SIMPSON, (RETIRED PRESIDENT, S.M. SIMPSON LTD.)

Troi Caldwell, President, IWA Local 1-423 and Jack Munro prior to his retirement in 1991 from the position of President, IWA Canada.

The Unions

"I don't regret a day of my work. I love the forest industry. The only regret I have is that a big part of my job has been in offices and on planes. I'd rather be out planting or pruning trees. Roaming around the forest for six months a year – that would be the perfect job."

– Wayne Nowlin (President, IWA Local 1-405).

"I've always been pleased to be a part of the interior industry. It has become pretty efficient over the years and almost every worker has some kind of specialized skill. Even though I've spent years in Vancouver with the IWA's national office, I'm still a champion of the southern interior. I think there are some good progressive people there who have a pretty good grasp of the problems and do a lot of things right. The southern interior isn't where I live now, but I have my roots there and still consider it my home."

– Jack Munro (Retired President, IWA Canada)

"Some highlights of my involvement over the years were the unsuccessful attempt to take over the IWA by the communist-led "Woodworkers Industrial Union of Canada" in 1948, the successful 7-month strike of 1967 and 1968, the implementation of a pension plan for all members in 1972 and the breaking away from an international union to form an all-Canadian IWA."

– Dell Welder (Retired Financial Secretary, IWA Local 1-423)

Wood products form the bridge, the unifying factor that binds the economic future of forest companies with forest unions, and unions with companies. Wood products needed in every corner of the world –

structural lumber, plywood, pulp, paper and thousands of other value added products – tie the producers with their employees. Together they experience economic turns and cyclical markets. Revenue from the manufacture of forest products must ultimately stretch to provide companies with operating expenses, wages, and hopefully, a return for capital invested.

While the potential for conflict between employers and unions is always present, there is a considerable effort made by both to find workable solutions to the issues of the workplace.

Within the forest industry the predominant union presence has been the IWA CANADA. As do many industrial and trade unions, the IWA traces its history back to the years of the Great Depression. The first IWA convention was held in Tacoma, Washington in July, 1937. Later that year, the B.C. district council of the IWA held its first meeting in Victoria.

The first Southern Interior IWA charter was granted in August, 1944 to Local 1-405 (Cranbrook). The founding documents bear the names of officers V. OAKLEY, A.S. MARIE and WILLIAM LANGMEADE. Charters were granted the following year to IWA Locals in Salmon Arm and Kelowna.

Other major unions representing interior forest workers include the COMMUNICATION, ENERGY & PAPERWORKERS' UNION (formerly the Canadian Paperworkers' Union) and the PULP, PAPER AND WOODWORKERS' OF CANADA (PPWC).

The complex nature of the forest industry calls for frequent interaction between forest companies and forest worker unions, as well as with non-forest unions. At various stages of the business, teamsters, stevedores, rail workers, government and construction workers, all play a role in getting wood products to market.

The IWA's first years were fraught with struggles over self-government by the union membership. External forces, some strongly left wing, were sometimes perceived as driving their political agendas through union ranks. Most union members were more interested in down-to-earth issues. Wages and working conditions topped their agendas.

To their credit, they remained determined, even through periods of severe tension and conflict, to maintain democratic decision-making within the union structure. The IWA leaders also upheld and worked toward the principle of Canadian autonomy.

Difficult internal conflicts also had to be dealt with during the formative years of the IWA. Splinter groups and pressure blocs formed within the structure, and varied from one region to another. Unions were fractionalized, for example, over the controversial decision to link with a national Canadian political party. In the late-40s the forest union aligned with the CANADIAN COMMONWEALTH FEDERATION (CCF), the forerunner of the New Democratic Party (NDP).

Through the 1940s, union factions struggling for power faced each other in court. Long, exhaustive legal battles drained the energy and resources of the combating factions. Membership control of the union was a key to settling specific disagreements and conflicts. By January of 1950, fractures appeared to have been healed.

In the early 1950s, there were moves to strengthen and consolidate the union presence in the interior. Although the process did not always run smoothly, setbacks and difficulties were eventually overcome. For many years now, the IWA has been the principal worker voice for the B.C. Interior forest industry.

The most memorable strikes in the Southern Interior occurred in 1953/54, 1967/68, 1981 and 1986. Most people involved share unpleasant memories of these strikes, as they were painful to everyone involved. The Southern Interior strike of 1967/68 was the longest in the history of the IWA. Approximately 4,000 IWA members struck southern interior operations on October 4, 1967. Work did not resume until seven months later, on May 13, 1968. The union gained base rate parity with the coast. It also moved closer to achieving a common contract expiry date with locals on the coast and in the northern interior. The positions, counter-positions and strategies of the 1967 strike were still topics of great debate in the years that followed the "winter of discontent".

All but a few ILMA member mills are now unionized. About 95 percent of logging for interior mills is performed by non-union contractors.

The longest serving President of IWA-Canada, JACK MUNRO, was one of Canada's more colourful union bosses. Originally from Alberta, "Union Jack" started with the interior forest industry as a welder for KOOTENAY FOREST PRODUCTS in Nelson. He was elected head of the local union in Nelson, and for a short time in 1968 was President of Local 1-405 Cranbrook. Munro's penchant for straight-talk earned him a rapid ascent through union ranks. By 1983 he was President of the IWA Regional Council No. 1. He retired from the IWA in 1991 and is now chairman of the FOREST ALLIANCE OF B.C.

"I'll never forget my first union assignment. I was given a second-hand brief case and a car and sent to a hornet's nest to try to settle differences. No one would even talk with me. I found a note on my car telling me to get out of town or else. I went to the town bar where I knew I'd meet the workers after their shift. After sitting through a big argument about who was going to beat me up, I bought them all a round of beer and walked out. After that we could talk."

– BILL "WILLIE" SCHUMAKER (RETIRED PRESIDENT, IWA LOCAL 1-423)

"I'm proud that we have the highest-paid millworkers in the world. It used to be that a sawmill job was considered pretty low on the totem pole and a last resort. I've seen the turnaround. Forest industry workers are top dog and a sawmill job is first now. My heart has always been with forestry. It's one hell of an industry. It's provided a lot of people in this province with a good living and it should continue to do that."

– BILL "WILLIE" SCHUMAKER (RETIRED PRESIDENT, IWA LOCAL 1-423)

Bill "Willie" Schumaker (on left) and Wayne Nowlin, on the occasion of Willie Schumaker's retirement, May, 1990.

"From 1956 to 1959 I was involved in labour negotiations for the ILMA. I have never thought of myself as a "saintly" individual but apparently Joe Morris, who was IWA chief in 1958, thought I was. In a write-up in the Penticton Herald, Joe likened me to St. George, riding a white horse, slaying the dragon (the IWA) and defending and protecting the "innocent" members of the ILMA. Labour relations were always difficult. I was certainly introduced to some interesting individuals and situations. In 1959, the Interior Forest Labour Relations Association was formed which removed labour negotiations from ILMA activities."

– AL MACDONALD, (RETIRED ILMA MANAGER)

6 year old C.F.I. plantation at Skelly Creek, northeast of Creston (spruce planted, pine is natural regeneration).

The Forest Service in the Southern Interior

"I was a Forest Ranger for the best part of 20 years. Everything I lived through as a Forest Ranger is now changed quite a bit. Before re-organization (of the Forest Service) the Ranger had a lot of discretionary powers. He lived in the community and knew the people involved – got to know them really well. You knew the ones you could trust and the ones you couldn't. It is very different now. Things have become very complicated. The system involves a lot of paper. I really yearn for the old days when you could deal with a fellow out on the stump and make some arrangements on how to log a ridge and do something different next to a stream without having to resort to a lot of registered letters and warnings and threats and discussions in the press and everything else."

– Bill Uphill, (Retired Vice President, Pope & Talbot Ltd.)

"In the days of the local Forest Ranger, a lot of the arrangements were on good faith. There were contracts, but there would be things not included, and if the Ranger wanted you to do them you just did. I think one of the things that helped was that the Ranger was part of the community too. And there was more trust that everyone would be fair and responsible and do their job properly. Where a bad operator disregarded those expectations, the Ranger took necessary action. History shows that bad operators didn't last long in the business."

– Gerald Raboch, (Chief Operating Officer, Riverside Forest Products Ltd.)

The people of British Columbia are the owners of 94% of British Columbia's commercial forest land base. Forest companies are tenant harvesters of the forest crop, and the provincial Forest Service is the agent for the landowners.

Tenancy agreements, administered by the Forest Service on behalf of the landowners, have evolved and matured gradually over the years. Forest management stages in B.C have accompanied attitudinal changes in society.

Prior to the Second World War, chief responsibilities of Forest Service departments were sharply defined. They made timber available for the forest industry, which in turn employed people and supplied wood products to other emerging industries. They collected forest revenue for the government, necessary for the young province to build an infrastructure. And they fought forest fires.

Very little technical management was done in the forests of the 20s and 30s. Parts of B.C. were logged for nearly half a century without much knowledge of forest inventory. Only a few far-sighted foresters and forest companies, mainly on the coast, took the initiative to establish small pioneer tree plantations.

Bountiful and lavish, forests in Canada's westernmost province were considered inexhaustible. Trees filled valleys and covered hills and mountainsides, for literally hundreds of miles at a stretch. [Approximately one half of the province's 95 million hectares is forested. Of the forested area, only about half is suitable for commercial or "working forest".]

Given so vast an area and so plentiful a resource, the province, in the early years, did not take serious notice of the cumulative effects of logging. For the first 50 or 60 years, management grappled mainly with the physically challenging but straightforward tasks of building roads and harvesting timber. In about the mid-30s, both government and industry began to awaken to the limits of the forest resource.

By the end of the 40s, it had become evident that more attention had to be paid to the dimensions and the conditions of the forests. Two government commissions, both headed by CHIEF JUSTICE GORDON SLOAN, marked the end of pioneer forestry, and the beginning of a new era.

The first Sloan Commission, in 1943-45, publicly examined forest practices in B.C. and was severely critical of many common forest practices of the time. It made forceful recommendations about changes to timber allocation, reduction of waste, reforestation and other aspects of forest management. These recommendations, and those of the second Sloan Commission ten years later, overhauled B.C. forestry from the roots up.

If government changes to policy and regulations were unpopular, Forest Service field personnel often bore the brunt of public dissatisfaction. The open bidding system, in place to award cutting rights to the highest bidder, created so many problems, for example, that the second Sloan commission recommended its replacement with the quota system.

Until roughly a decade ago, keeping up-to-date with reforestation was not a very determined commitment of any government in B.C. "Permanent" silviculture funds set up under various administrations each lasted no more than a few years. Funds earmarked for silviculture were tapped to pay for other government programs whenever the economy took a downturn. The area of forest land classified as "NOT SATISFACTORILY RESTOCKED" (NSR) consequently did not shrink, but grew ever larger. Some NSR had been created by wildfires and insect kills. But by far, most of the NSR had been logged over and not planted back.

> *"A change for the better is the amount of silviculture going on. Over the years the Forest Service had various planting funds. Each time things got tough the funds were transferred by government to another priority. We had a slogan in the mid-1970s. It was "75 in '75". We were going to try to plant 75 million trees in 1975. We only made 60 million that year. Last year we planted something like 250 million trees. We are going from basic silviculture to intensive silviculture. We can now start to tend what we have – nurse it and look after it better."*
>
> – ED DORION, (PRESIDENT & CEO, EVANS FOREST PRODUCTS LIMITED)

Sustained yield objectives were adopted as permanent forest policy, shortly after the second Royal Commission. One obvious requirement of sustained yield, strongly emphasized in Sloan's 1956 report, was that logged forest land be regenerated as soon as possible after logging. Greater emphasis was placed on prompt planting of sites wherever prospects for natural regeneration were in question.

Natural regeneration has restocked many logged and burned forests quickly and well. After all, natural regeneration has been occurring after bug kills and huge fires for thousands of years. In the case of the immense lodgepole pine resource, naturally regenerated stands often have too many trees per hectare – a condition foresters refer to as "XSR". Foresters usually need to thin these naturally regenerated young trees to give the others room to grow. Twenty years ago thinning and spacing programs were rare. They are now a major part of stand treatment in many interior forests.

On some of the best growing sites in the province, however, naturally regenerated forests grew back in deciduous brush, with few conifers. In many of these cases, expensive brushing and weeding programs have been needed to get a commercial crop re-established. No one treatment can be expected to work everywhere.

It soon became obvious that on many forest sites reforestation would have to follow logging in order to ensure availability of a second growth forest. The first major replanting programs began in the 1930s, but efforts remained sporadic until well after the Second World War.

Today, reforestation to a "free growing" standard is an ongoing legal obligation. Where natural regeneration will not succeed, planting is conducted. As part of the reforestation requirement, seed is collected from suitable species in the right locations and held in cold storage until required. Seedlings are raised in nurseries and, at the appropriate time, planted in the harvested areas. The seedling survival rate in the B.C. interior is currently higher than 80 percent. For some time, B.C.'s reforestation program has had its own seed orchards, as well as the so-called "plus tree" initiative that selects seed from genetically superior trees.

Some of the interior's early first pass logging has regenerated naturally into productive second growth stands. Some of these forests are being logged now and many more are nearing the age when second pass logging can be undertaken.

In response to pressure from industry, local government and a concerned public, senior governments finally began the task of catching up on reforestation and reducing the backlog of NSR. The federal/provincial Forest Resource Development Agreement (FRDA) undertook the challenge of reducing this backlog of NSR. Current Chief Forester John Cuthbert has reported that there is still a small area of backlog, but B.C. forest regeneration is now current with the harvesting rate. Some old NSR still remains, a legacy of the inadequate reforestation practices of the past. Catch-up planting is quickly diminishing the size of NSR land.

The level of forest management has varied with changes in government policy, and as well, with changing governments. Increasingly, governments have shifted the responsibility for forest management to industry. In 1987, the "basic forestry" program (the establishment of a new crop of trees of sufficient numbers per hectare) has been transferred from government to industry. Forest companies are now responsible for reforestation, along with a growing list of other obligations.

"The most important change is the planting of more trees – regeneration of the forest. That is the biggest and best change as far as I'm concerned."

– TARA GHOG, (OWNER, ASPEN PLANERS LTD.)

"The impression the environmentalists have is that you go out there and log and you destroy the land. Now, people are much more aware of the importance of the industry and know they are going to have to restock the forest to create another crop. I feel that most people are not anti-logging but they are against logging and leaving it NSR (not sufficiently restocked)."

– ED DORION, (PRESIDENT & CEO, EVANS FOREST PRODUCTS LIMITED)

Sweeping changes in B.C. forest policy were also ushered in by the 1976 Royal Commission into forestry, headed by Dr. Peter Pearse. Pearse Commission recommendations resulted in a new Forest Act and amendments in related regulation.

Community-based Forest Rangers and staff had, by long tradition, become part of the fabric of small towns and settlements across the province. The personal contact between Forest Service people and the local population had made for open communication and dialogue.

In the late 1970s and early 1980s, the government re-organized the Forest Service, from the top down. Under the new name, Ministry of Forests, government forestry employees were centralized into a few District headquarters. Local ranger stations in many small centers throughout the province were closed and the traditional position of Forest Ranger was eliminated. Stations that had been integral parts of the community were sold off to gain efficiencies in cost and administration. The centralization move removed the grassroots contact of the government forest staff within these communities.

The re-organizational decision was contentious at the time, and remains controversial. The merits and drawbacks of centralization are still being debated. Whether the new Ministry of Forests is more effective than the former Forest Service is arguable. One consequence of the action has, however, been clear. Centralization effectively de-personalized forest management and led to a loss of community input and support.

> *"The rate that changes are occurring is what strikes me the most. It used to be that something would stay in place for a couple of years. But it now seems that you don't really have a chance to get used to something happening before it changes again. The rate of change has really accelerated. Now it seems the only thing that is going to remain the same is more and more change."*
>
> – ABE FRIESEN, (PRESIDENT, POPE & TALBOT LTD.)

Major changes in the Forest Act called for a much more detailed inventory of forest and rangeland and a continuously updated five-year plan for the operations and management of these resources.

While the Ministry was now required to intensify its scrutiny of forestry activities, its massive job of forest protection still had to be carried on. From a network of manned lookout stations that scanned the countryside watching for fires, forest fire detection and suppression has developed into space-age technology. Now satellite monitoring systems record not just every fire, but virtually every lightning strike. Lightning prone areas are monitored more closely, and the speed and efficiency of helicopter transport and fixed-wing water-bombers have made initial attack and suppression of fires increasingly effective. The province gets as many lightning strikes as ever, but far fewer develop into major fires.

Among the current research projects of forest scientists are studies that deal with forest insects and diseases. Huge insect outbreaks, especially bark beetle epidemics, have devalued thousands of trees each year and have caused mortality across literally miles of mature forest.

The mandate of the Ministry of Forests has evolved – from primarily fire-fighting, allocating timber and collecting government revenue – to include the mediating of an intense public debate over forestry and environmental issues. The Ministry's broader spectrum of responsibilities requires it to manage the forest for a growing list of non-timber, non-commercial values.

Over the past decade, water, viewscapes, wildlife and fish, parkland, wilderness and many other non-timber values have been built into forest management policies and practices. The Ministry of Forests, meanwhile, must continually strive to provide sufficient timber to keep forestry communities viable.

It has fallen upon the Ministry to try to integrate the varied uses and to accommodate advocates for often conflicting land uses. Ministry of Forests and forest industry employees have found it necessary to communicate with and relate to citizen groups or individuals with concerns or outright opposition to certain forest practices. Often these are people with no historic contact with either the industry or the forest itself. Ministry staff have discovered that a new set of mediation skills is required to cope with this phenomenon. The learning experience is ongoing.

The government's introduction of a legislated provincial "Forest Practices Code", presently in the design stage, has long been supported by the B.C. forest industry. A key element of the new code will require independent auditing of forest practices. The forest industry of the southern interior pins great hope on this process, as a key to restoring public confidence in forest land management.

Harrop Forest Nursery, Nelson, B.C., Pacific Reforestation Technologies, Englemann Spruce, 1 year old

Reforestation in progress in the Bush River drainage on blocks recently harvested by Evans

The Future

A New Working Environment

The evolution of the forest industry in the Southern Interior of B.C. is an unfinished story. From the Second World War onward, enormous changes have transformed every aspect of the forestry business.

A high degree of adaptability and resourcefulness on the part of individuals and corporations has enabled forestry to survive through these changes. The scope and rate of change can be expected to accelerate in the future.

Despite the arduous physical challenges of early forestry, pioneer days seem simple by comparison to the stresses and complexities of today's environmental, economic and social demands on the forest land base.

In the early days the viability of the forest sector was limited mainly by the mostly physical challenges of finding markets and getting forest products delivered to them. Then, as now, certain undeniable economic realities dictated the kind of industry that could exist, how it would perform and ultimately, whether a company in a given region was sustainable or not. The long list of companies that have passed into history is evidence of the risks and changing nature of the industry, and the inescapable need to pay close attention to the bottom line.

Original challenges have not disappeared, while a myriad of new challenges have been added. As the industry has become more visible, and public interest in forestry has reached across the entire province, and beyond, complexities have multiplied. Pioneer lumbermen might not even recognize forestry in the 1990s as the same business.

Some observers of the history and evolution of the industry worldwide, have noted that forestry passes through four general stages of development. Countries that have logged for centuries, and are on their fifth or sixth harvest on the same sites, sometimes took many centuries to progress through the stages. After a century of logging, and still mainly harvesting previously unlogged stands, B.C. forestry has evolved through these stages in a compressed time frame.

Four Stages of Forestry in B.C.

Stage 1.	**Unregulated exploitation**	Before the Second World War
Stage 2.	**Administrative approach:** The beginning of "forestry" as the risk of "exploitative" phase is recognized. Policies established which were applied uniformly to wide areas.	1950s to 1970s
Stage 3.	**Ecological approach:** "Site specific" prescriptions emerge as the broadly applied policies of the "administrative phase" are found to work well in some areas and not in others.	1980s
Stage 4.	**Social forestry:** Interests such as aesthetics and wilderness values are increasingly included in the planning processes.	1990s and future

The inclusion of ecological and social factors into forest decision-making has factored more than trees, game animals and fish into the management equation. The term, "biological diversity", is everyday language wherever logging plans are discussed and designed.

Sweeping changes in forest policy during the 80s and early 90s coincided with a dramatic rise in public interest in the overall environment. It now seems that every region of the province has its own set of land use conflicts over parks, wilderness, wildlife, fisheries, watersheds, viewscapes and other forest values. Responsible advocates on all sides of the land-use debate became acutely aware that new systems of land allocation and land use were needed to avoid and resolve the increasing number of valley-by-valley conflicts.

Some high profile examples of conflict have grown into bitter local debates and deteriorated into illegal road blockades, court injunctions, arrests and jail terms for protesters. These emotionally charged conflicts have been socially disruptive in more than a few cases. Entire communities have been divided over the worst of them.

"In the old days we had more time. We were proud of what we did. There was more self-esteem. Now some people who don't work in the forest industry tend to treat you like a second class citizen."

– DAVE MCINNES, (CHAIRMAN OF THE BOARD, WEYERHAEUSER CANADA LTD.)

"Loggers can no longer just go and cut trees and haul them to the mill. Now they have to really plan their roads, plan their logging and they have to know how they are going to replant and how they can tend the stand afterward. It makes people take a look at the whole picture and be rewarded for their stewardship of the resource. This is a change for the better. I just hope the balance doesn't shift too far the other way, but there are some signs of that."

– FRED STINSON, (RETIRED, ILMA PRESIDENT)

"I see a growing awareness that the trees belong to the community and there is an expectation that the community wants a better say in how these trees are managed and where. We have communities where people are fighting each other. In an overview this might be a good direction that society is taking. If intelligently applied it can be a good thing. But in the extreme it can be an abusive thing."

– IKE BARBER, (PRESIDENT, CHAIRMAN & CEO, SLOCAN GROUP)

Many varied initiatives have been attempted to search for common ground over matters of forest land use. The land use discussion in the southern interior has become much more broadly-based than anything experienced in the past. In addition to the current Kootenay/Boundary CORE (Commission on Resources and Environment) process, in some interior areas, stakeholders from industry and the concerned public have developed consensus-based working groups, such as local Round Tables, offshoots of the B.C. Round Table on the Environment and the Economy.

Hopefully, these consensus building processes will, in the long run, build trust and lead to broader public understanding and respect for separate positions.

On all fronts, provincial, national and global economies are striving to keep pace with change. From outside the forest industry, some are contending that B.C. should make a transition, from its traditional resource-driven economy, to a service-oriented economy.

The service sector is indeed gaining ground as a contributor to the provincial economy.

Our society cannot, however, afford to give up on the wealth generated by forestry, and other resource-based industries. Recent conclusions of the FOREST RESOURCES COMMISSION and the B.C. Round Table verify that the extraction and export of natural resource products still drives the B.C. economy. And forestry is the number one engine.

Even though the largest sector of the provincial economy is now service-oriented, much of the service sector exists because of the forest industry, and other resource-based sectors. Resource industries and the supply and service sector are heavily and mutually inter-dependent.

Based on Ministry of Finance analysis, the B.C. Forest Resources Commission issued a press release in March, 1992 indicating that forestry remains the province's dominant employer. It provides more than 50 percent of all "basic" jobs in 25 of the province's regional areas, which contain more than 200 separate communities.

The Forest Resources Commission cautioned: "Great care must be taken when considering actions that may result in a significant reduction in forest industry activity... Failure to do so could result in major social dislocation and community upheaval outside of the Lower Mainland."

The vitality of B.C.'s tourism industry is also related to resource extraction. Layoffs or slow-downs in the resource sectors have been shown to result in declines in tourism. Travel and recreational activities are, necessarily, one of the first curtailments of out-of-work families.

There is potential for the expanding tourism industry to contribute to an increasing portion of the provincial economy, especially within the B.C. interior. In discussions about the relative merits of tourism and forestry, however, there is a tendency to present an either/or scenario. Save for some site-specific situations, tourism and forestry need not be mutually exclusive. As fiscal burdens on government continue to grow heavier, the province will need to secure and increase the economic benefits that derive from both tourism and forestry.

As the generator of more new wealth than any other industrial activity in the province, forest management will become an increasingly vital priority in striving for economic, social and environmental sustainability of communities province-wide.

In an industry that has already been marked by enormous change, future forestry in the southern interior, as elsewhere in the province, will be required to display more flexibility and adaptability. These qualities, characteristic of B.C. forestry from the beginning, have helped the industry endure through much difficulty and many changes. Forestry will be tested even more in the future, as forest policies and practices will most certainly continue to evolve in concert with public attitudes, and social, economic and environmental expectations.

"I think it is positive that the public is taking an interest in the resource. We have such a value in our forests. Over the years we may not have been as careful as we might have been. It is a good thing for us to be more considerate of the resource and try to make the most of it and utilize it for multiple use. We also have to use it for more higher value products. That is a positive trend."

– ABE FRIESEN, (PRESIDENT, POPE & TALBOT LTD.)

"There is no doubt that the 1990s will be remembered as the decade of the environmentalists. The industry has made a lot of mistakes and now it is a case of communication with environmental groups. The negative side has been told, but not the positive side. We have to inform the public about the industry and discuss, not only environmental values, but also the importance of the industry to the economy."

– GORDON STEELE, (PRESIDENT & CEO, RIVERSIDE FOREST PRODUCTS LTD.)

"Today it is harder to be as optimistic about the future of our country as my father was. Look at the debt we have and things going the way they are. Look at the mining industry and see what's happening. We may see the forest industry move on too, although I certainly hope not."

– JACK WIGEN, (PRESIDENT, WYNNDEL BOX & LUMBER LTD.)

"Recently I was up in the Trout Lake area and looked at the area around Shelter Bay (on the Arrow Lakes). I personally had logged there in the 1960s and it was so good to see how well it had come back. It just looked super – a healthy excellent, second-growth stand."

– DAVE MCINNES, (CHAIRMAN OF THE BOARD, WEYERHAEUSER CANADA LTD.)

"We were more aware of what went on around us during the days of crosscut saws and horse logging. Then we entered a time when the economies of scale started to take over as we struggled to be competitive in world markets. The hustle and bustle and noisy equipment made us lose contact with what was going on for a while. We now know we can remain competitive in world markets. The up-side of environmental issues is that we regained something important – the desire to get back in closer touch with the land."

– GERALD RABOCH, (CHIEF OPERATING OFFICER, RIVERSIDE FOREST PRODUCTS LTD.)

From the President's Chair:

Forestry's evolution has been a drama about people – men and women and families who have faced challenge with energy and creativity. Their hard work, in forests that belong to all the people of B.C., and in scores of remote sawmills, provided the wood products that literally built the Interior. The dynamic story of the Southern Interior forest industry is still being lived out daily.

The nostalgic look back at Interior history has been a 'how-to' lesson in meeting challenge. With lessons from the past in mind, the industry shifts its interest to challenges of the future.

Forestry will continue to adapt to a world that is changing at an accelerated pace. A major hurdle in the course ahead will be the creation of provincial and regional land use strategies that sustain a strong forest industry and a healthy environment. Redoubled effort will need to be applied to maintain allowable annual cuts at levels that stabilize timber-providing communities. Forest companies will also be challenged, as social, environmental and economic pressures grow heavier and more complex, to venture into more intensive forest management and to seek new ways to create more B.C. jobs with B.C. wood.

We hope "Loggers and Lumbermen" has offered insight into the enterprising spirit of people who were committed and dedicated to a forest-based way of life. The world of forestry near the beginning of the 21st century is vastly different from at its start, but the most important part of the industry is unchanged. Men and women with an enduring respect for the land are, and ever will be, a strength of the Southern Interior.

JOHN MARRITT, PRESIDENT, ILMA

Lower Lussier River southeast Canal Flats – logged to control spread of pine beetle – natural regeneration.

Before – Drag scarified in 1979 for natural regeneration

After – 1990 – Excellent results after 11 years.

Index of People's Names

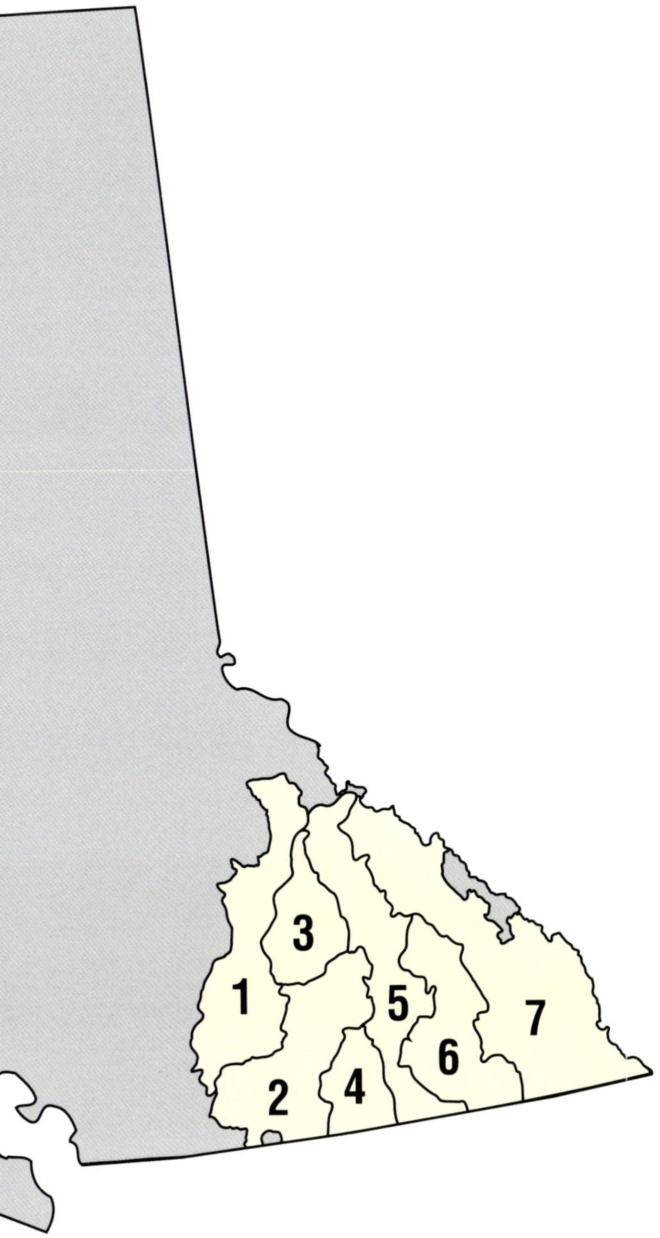
1
2
3
4
5
6
7

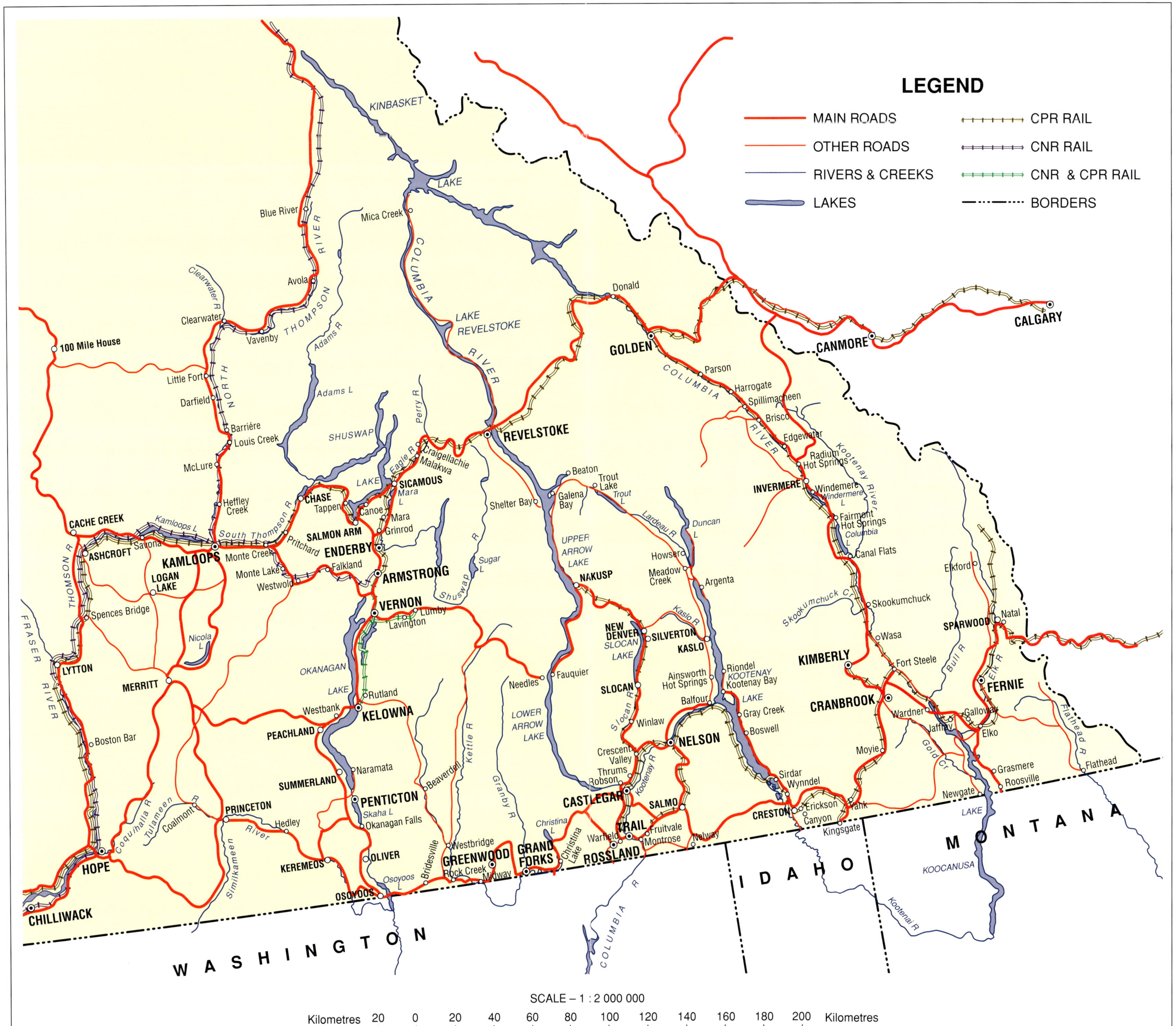
LEGEND
MAIN ROADS
OTHER ROADS
RIVERS & CREEKS
LAKES
CPR RAIL
CNR RAIL
CNR & CPR RAIL
BORDERS
KINBASKET
LAKE
Mica Creek
Blue River
Avola
Clearwater R
Clearwater
Vavenby
NORTH THOMPSON RIVER
100 Mile House
Little Fort
Darfield
Barrière
Louis Creek
McLure
Heffley Creek
Adams R
Adams L
SHUSWAP
LAKE
COLUMBIA
LAKE REVELSTOKE
RIVER
Donald
GOLDEN
CANMORE
CALGARY
Parson
Harrogate
Spillimacheen
Brisco
Edgewater
Radium Hot Springs
REVELSTOKE
Perry R
Eagle R
Craigellachie
Malakwa
SICAMOUS
Mara L
CHASE
Tappen
Canoe
Mara
Grinrod
SALMON ARM
ENDERBY
CACHE CREEK
Kamloops L
South Thompson R
Pritchard
ASHCROFT
Savona
KAMLOOPS
Monte Creek
Monte Lake
Westwold
Falkland
ARMSTRONG
LOGAN LAKE
THOMSON R
Spences Bridge
Nicola L
VERNON
Lumby
Lavington
Shuswap R
Sugar L
FRASER RIVER
LYTTON
MERRITT
OKANAGAN LAKE
Rutland
Westbank
KELOWNA
PEACHLAND
Boston Bar
Naramata
SUMMERLAND
PENTICTON
Skaha L
Okanagan Falls
Beaverdell
Kettle R
Granby R
Coquihalla R
Tulameen R
Coalmont
PRINCETON
Hedley
Similkameen River
HOPE
CHILLIWACK
KEREMEOS
OLIVER
Osoyoos L
OSOYOOS
Bridesville
Westbridge
GREENWOOD
Rock Creek
Midway
GRAND FORKS
Christina L
Christina Lake
Beaton
Trout Lake
Trout L
Galena Bay
Shelter Bay
Lardeau R
Duncan L
UPPER ARROW LAKE
NAKUSP
Howser
Meadow Creek
Argenta
Kaslo R
NEW DENVER
SLOCAN LAKE
SILVERTON
KASLO
Fauquier
Needles
SLOCAN
Ainsworth Hot Springs
Riondel
KOOTENAY Kootenay Bay
LAKE
Balfour
Gray Creek
Boswell
LOWER ARROW LAKE
Slocan R
Winlaw
Crescent Valley
Thrums
Robson
NELSON
Kootenay R
CASTLEGAR
SALMO
TRAIL
Warfield
Fruitvale
Montrose
ROSSLAND
Nelway
Sirdar
Wynndel
CRESTON
Erickson
Canyon
Kingsgate
Yahk
INVERMERE
Windemere
Windermere L
Kootenay River
Fairmont Hot Springs
Columbia L
Canal Flats
Skookumchuck C
Skookumchuck
Wasa
Fort Steele
KIMBERLY
CRANBROOK
Wardner
Jaffray
Moyie
Gold Cr
Elkford
SPARWOOD
Natal
Bull R
Elk R
FERNIE
Galloway
Elko
Grasmere
Roosville
Flathead R
Flathead
Newgate
LAKE KOOCANUSA
Kootenai R
WASHINGTON
IDAHO
MONTANA
COLUMBIA R
SCALE – 1 : 2 000 000
Kilometres 20 0 20 40 60 80 100 120 140 160 180 200 Kilometres